Nanoscale Science and Engineering Forum 2018

Core Programming Area at the 2018 AIChE Annual Meeting

Pittsburgh, Pennsylvania, USA
28 October - 2 November 2018

ISBN: 978-1-5108-7614-9

Printed from e-media with permission by:

Curran Associates, Inc.
57 Morehouse Lane
Red Hook, NY 12571

Printed by Curran Associates, Inc. (2019)

For permission requests, please contact AIChE
at the address below.

AIChE
120 Wall Street, FL 23
New York, NY 10005-4020

Phone: (800) 242-4363
Fax: (203) 775-5177

www.aiche.org

Additional copies of this publication are available from:

Curran Associates, Inc.
57 Morehouse Lane
Red Hook, NY 12571 USA
Phone: 845-758-0400
Fax: 845-758-2633
Email: curran@proceedings.com
Web: www.proceedings.com

TABLE OF CONTENTS

Some abstracts in this book refer to images that were not made available by the publisher

540534 Scalable Fabrication of High Performance Microbatteries, Biosensors and Optical Elements Via Nanoimprinting of 3-D Metal Oxide Structures

Abstract:

We have developed an approach for the direct imprint patterning of crystalline metal oxides using a soft polymer master and inks containing high concentrations of crystalline nanoparticles dispersed in solvent and/or in sol-gel precursors to a desired inorganic phase. High aspect ratio nanostructures and sub-100 nm features are easily realized. Residual layer free direct imprinting (no etching required) is achieved by choosing an ink with the appropriate surface energy to ensure dewetting at stamp-substrate interface. The technique is further extended to stack the nanostructures by deploying a layer-by-layer imprint strategy. The method is scalable and can produce large-area, device-quality nanostructures in a rapid fashion at low cost. Here we illustrate the utility of this direct patterning technique by the fabrication of fully-integrated, high-performance lithium-ion microbatteries, high-sensitivity biosensors, and advanced optical elements.

525119 Enhancing the Selectivity of Gas Sensors By Pre-Separation with Membranes or Powder Filters

Abstract:

A challenge in gas sensing (e.g. breath analysis or indoor air quality monitoring) is the accurate detection of trace-level gases in complex mixtures. While modern chemoresistive gas sensors can be extremely compact, inexpensive and highly sensitive, their success is still limited by *selectivity*. Here, we dramatically enhance their selectivity by pre-separating gas mixtures with *zeolite membranes* or *activated alumina powder* for detection of formaldehyde or isoprene at the ppb-level at 90% relative humidity. Formaldehyde is a potential breath marker for lung cancer and a tracer for indoor air quality monitoring while isoprene is a versatile breath marker for non-invasive monitoring of high blood cholesterol levels as well as for influenza, end-stage renal disease, muscle activity, lung cancer and liver disease with advanced fibrosis. As proof-of-concept, a zeolite MFI/Al_2O_3 membrane is placed upstream a highly sensitive but weakly selective Pd-doped SnO_2 sensor. Their combination exhibits exceptional selectivity (*>100*) for formaldehyde down to *30 ppb* at 90% relative humidity, outperforming state-of-the-art detectors by more than an order of magnitude (1). Similarly, we present a pre-separation concept based on activated alumina powder that retains hydrophilic compounds (e.g. ketones, alcohols, ammonia) representing major interferants in breath while hydrophobic isoprene is not affected. This results in fast (10 s) measurement of isoprene down to 5 ppb at 90% relative humidity using the same sensor with outstanding selectivity (>100) to breath-relevant acetone, ammonia, ethanol and methanol, that is superior to state-of-the-art isoprene sensors. Most importantly, when exposed continuously to simulated breath mixtures (four analytes) for eight days, this filterâ€"sensor system showed stable performance (2). Both pre-separation systems can be interfaced readily with portable breath detectors or compact indoor air monitors.

1. A.T. Güntner, S. Abegg, K. Wegner, S.E. Pratsinis, Zeolite membranes for highly selective formaldehyde sensors, *Sensors and Actuators B: Chemical*, **257**, 916-923 (2018).
2. J. van den Broek, A.T. Güntner, S.E. Pratsinis, Highly Selective and Rapid Breath Isoprene Sensing Enabled by Activated Alumina Filter, *ACS Sensors,* **3**, 677-683 (2018).

538950 Low-Cost and High-Throughput Synthesis of Copper Nanopowder for Nanofluid Applications

Abstract:

Cu is low cost and efficient filler for nanofluids. However, commercially available Cu nanopowders are as costly as silver. A few methods are available for synthesis of Cu nanopowder, but the amount of powder produced by these methods are inadequate. Solution based high throughput continuous method for production of kgs of Cu nanopowder has been reported in this work. Cost of production is minimal in this method and nanoparticles are separated from the suspension by a controlled flocculation process. The proposed method relies on the higher solubility of cuprammonium complex in an aqueous medium and its chemical reduction using hydrazine hydrate. The colloid has been stabilized using sodium citrate and obtained as sub 50 nm powder re-dispersible in water and EG. Upon re-dispersion the sub 50 nm clusters have broken down to sub 10 nm particles. Continuous synthesis using a CSTR has been demonstrated, and the quality of particles are similar to that of the batch method. Stable nanofluids have been prepared by re-dispersing the nanopowder in water and EG. Good enhancement (~5%) in thermal conductivity for Cu-EG nanofluids and reasonable enhancement (~1%) for Cu-water nanofluid have been observed with mere 0.6 and 0.3 vol% loading respectively.

527598 Synthesis and Online Characterization of Metallic Nanoparticles By Spark Ablation

Abstract:

Aerosol synthesis is a facile route for the continuous production of nanoparticles. Among the many existing gas phase processes, inert gas condensation routes are outstanding as they can yield particularly small and pure nanoparticles. Compared to other methods like flame spray pyrolysis or CVD, no by-products, e.g. carbonaceous species, are formed. Among the various inert gas condensation processes, spark ablation is a facile method to produce small metallic nanoparticles by creating a high voltage discharge between two electrodes. This leads to heating and partial evaporation of the electrode tips and, after recondensation, the formation of nanoparticles. These can be used as seeds for heterogeneous nucleation processes or to study aerosol mechanics. Spark ablation already proved to be suitable both for the production of a number of different materials [1] as well as for scalability [2]. Within this work, particles made from a number of different metals are synthesized with controlled size.

The setup used in this work adapts the design used in [1] and comprises of a high current power supply, rotameters to adjust the carrier (He or Ar, 99.999 % purity) and quench gas (N_2, 99.999 % purity) flows and a sixfold vacuum flange. Exchangeable capacitors and resistors are integrated into the spark circuit. Thereby, both spark frequency and energy can be influenced. The electrode rods typically have a diameter of 2 mm and hemispherically shape tips and are held by brass adapters allowing for quick exchange. The produced aerosols are measured by a combination of a TSI 3080 electrostatic classifier equipped with a TSI 3077A ^{85}Kr neutralizer, a TSI 3085 Nano DMA and a TSI 3776 condensation particle counter.

In the present study, the influence of both process and material parameters on the produced particles was investigated. Therefore, eleven different elements were tested, namely Ag, Al, C, Cu, Hf, Mo, Nb, Sn, Zn, Ti and W. These metals represent a wide range of material properties - from very low (Sn, Zn) to very high melting (W, C) elements and from highly (Cu, Ag) to weakly conductive (Ti, C) ones. Furthermore, the circuit current and capacitance as well as the carrier gas flow and the electrode diameter were investigated for their influence on the produced particles' properties.

The circuit capacity C and current I turned out to be the most important process parameters to tune the size of the produced particles. A higher capacity leads to a higher dissipated energy per spark discharge and reduces the spark frequency at the same time. The fact that larger particles are produced when increasing C indicates a strong influence of energy dissipation along the electrodes and thus, their thermal conductivity as a change in circuit capacitance has no influence on the time-averaged energy input into system. With higher capacitance, an increasing amount of the energy from the spark discharge cannot be dissipated be means of thermal conduction and is instead used for heating and evaporation of electrode material. An increased circuit current also leads to larger particles; the dependency can be fitted best with a power law function which further supports the hypothesis of the strong influence of heat dissipation. Consequently, in an approach to nondimensionalize the problem, a modified Fourier number was defined that also includes the melting and boiling point as well as the enthalpies of fusion and evaporation of the respective elements. Correlating the mass yield of different electrode materials against their modified Fourier number rat otherwise the same process conditions showed an inversely proportional dependency for the majority of elements [3].

Summarizing, an overview of the most important process parameters for spark ablation synthesis and their influence on the produced particles' properties is given. It is pointed out that heat dissipation along the electrodes is a crucial phenomenon when it comes to correlating material properties of different elements with the mass yield achieved by using the respective electrode material.

This work was supported by the German Research Council (DFG) and the Cluster of Excellence "Engineering of Advanced Materials" (EAM).

[1] N. S. Tabrizi, M. Ullmann, V. A. Vons, U. Lafont, and A. Schmidt-Ott, "Generation of nanoparticles by spark discharge," *J. Nanoparticle Res.*, vol. 11, no. 2, pp. 315–332, Feb. 2009.

[2] J. Feng, G. Biskos, and A. Schmidt-Ott, "Toward industrial scale synthesis of ultrapure singlet nanoparticles with controllable sizes in a continuous gas-phase process," *Sci. Rep.*, vol. 5, no. 1, p. 15788, Dec. 2015.

[3] M. Domaschke, M. Schmidt and W. Peukert, "Aerosol synthesis of ultrafine monometallic nanoparticles by spark ablation" J Aerosol Sci., *submitted*

527759 Gas Phase Coating of Germanium Nanoparticles with Silicon

Abstract:

Nanocrystalline silicon (Si) and germanium (Ge) are promising materials for the application in printable electronics (Holman et al., 2010). Nanoparticles (NPs) of these materials can be dispersed in organic liquids and deposited as thin functional films. Especially hybrid structures made of both materials provide a great opportunity to manipulate the material properties. In previous works we have already shown the coating of Si NPs with Ge to produce defined patches on the particles (Mehringer et al., 2014). In this work we report our progress in coating Ge NPs with silicon. A major problem of Ge NPs is the high surface reactivity. Ge oxidizes very quickly and the oxide has a poor chemical stability. GeO_2 e.g. is highly dissolvable in water in contrast to SiO_2. Therefore, passivation of Ge NPs with silicon is a promising strategy to ensure high device performances (Hunter et al. 2017).

For the experiments, a two stage hot-wall reactor (HWR) setup combined with an external seeding unit is used. This unit consists of a hot wire generator (HWG) producing small metallic seed particles which are essential for the growth of well-defined Ge NPs in the first stage (HWR I). The seed particle stream is directly mixed with the precursor monogermane and injected into HWR I. In the second stage (HWR II) monosilane is added for coating the Ge NPs with silicon. In the first stage, the temperature is kept constant at 800 °C while in stage 2 it is varied in the range of 600 1000 °C. At the reactor exit the aerosol is quenched by nitrogen. The particles sampled *via* a low-pressure impactor by deposition directly on substrates or *via* a filter.

The HWG offers the great advantage of precise control over the seed particle properties and thus over the final Ge NPs. The seed particles suppress the homogenous nucleation which would otherwise lead to high nucleation rates and small agglomerated particles. With the external seeding unit the spherical Ge NP synthesized in HWR I show a narrow size distribution (geometric standard deviation GSD < 1.1). The mean particle diameter can be varied in the range of 25 to 50 nm. These particles serve as cores for silicon coating in the second stage. The morphology of the coated particles is investigated by SEM or TEM images. On the germanium cores patches of silicon are visible instead a complete layer. This can be explained by energetically favoured sites for the initial island growth. Due to the slight crystal lattice misfit, the Si grows preferentially on itself and only partially around the Ge core. An increasing synthesis temperature leads to only one big patch because of the higher mobility of the adatoms. Therefore, the patch formation can be divided into two regimes. On the one hand a diffusion dominated regime at high temperatures and on the other hand a nucleation dominated regime at low temperatures. Furthermore, the patch number and size strongly depends on the Si concentration. A higher concentration leads to a decrease in the number of patches while simultaneously increasing the patch volume.

The progress on coating Ge NPs with silicon will be demonstrated, regarding the influence of the process parameters e.g. temperature, pressure, precursor concentration etc. Furthermore the growth mechanism will be discussed in detail.

This work was supported by the German Research Council (DFG) and the Cluster of Excellence "Engineering of Advanced Materials" (EAM).

C.Holman, Z. and Kortshagen, U.R. (2009). Solution-processed germanium nanocrystal thin films as materials for low-cost optical and electronic devices. *Langmuir,* 25(19):11883–11889.

Mehringer, C., Wagner, R., Jakuttis, T., Butz, B., Spiecker, E., and Peukert, W. (2014). Gas phase synthesis of anisotropic silicon germanium hybrid nanoparticles. *J. Aerosol Sci.*, 67:119–130.

Hunter, K.I., Held, J.T., Mkhoyan, K.A., and Kortshagen, U.R. (2017). Nonthermal Plasma Synthesis of Core/Shell Quantum Dots: Strained Ge/Si Nanocrystals. *ACS Appl. Mater. Interfaces,* acsami.6b16170.

534491 Mass Production of Nanoscale Materials with Uniform Ultralarge Mesopores Via Colloidal Solution Combustion Synthesis

Abstract:

Mass Production of Nanoscale Materials with Uniform Ultralarge Mesopores via Colloidal Solution Combustion Synthesis

Mesoporous metal oxides are commonly synthesized via soft or hard templates. However, these multistep laborious methods normally result in low product yield. Furthermore, mesoporous hard templates (e.g., SBA-15, KIT-6, and CMK-3) are very expensive, and high capital costs are needed for the production. Facile and economical synthesis of mesoporous oxides with tailored porosity on a large scale still remains a challenging task for commercialization.

Here we demonstrate a scalable, economic, energy and time efficient colloidal solution combustion synthesis (CSCS) method to produce crystalline mesoporous materials with tailored porosity using inexpensive colloidal silica [1,2]. CSCS utilizes the internal chemical energy of reactants which is released in the form of the heat, converting precursors into the products. For the proof of concept 0.5 kg of crystalline CeO_2 with uniform 22 nm ultralarge mesopores, 130 m^2/g surface area, 0.6 ml/g pore volume and high yield has been synthesized at low cost. Importantly, by eliminating the use of an organic solvent or surfactant, CSCS provides relatively green chemistry route to a scalable synthesis of nanomaterials. In addition to large-scale synthesis capability, CSCS method provides exceptional advantages compared to other well-established template-assisted methods. Specifically, the large surface area, uniform ultralarge mesoporosity, high crystallinity, and small crystal size of the oxide can be achieved at the same time, whereas prolonged high-temperature post-synthesis calcination for crystallization usually leads to the collapse of the mesoporous structure in other template-assisted methods. These can be achieved because the formation and crystallization of nuclei occur almost simultaneously through the heat generated from the exothermic reaction between the reactants in the CSCS.

[1] A. A. Voskanyan, K.-Y. Chan, C.-Y. V. Li, Colloidal Solution Combustion Synthesis: Toward mass production of a crystalline uniform mesoporous CeO_2 catalyst with tunable porosity. *Chem. Mater.* **2016**, *28*, 2768-2775.

[2] A. A. Voskanyan, K.-Y. Chan, Scalable Synthesis of Three-Dimensional Meso-Macroporous NiO with Uniform Ultralarge Randomly Packed Mesopores and High Catalytic Activity for Soot Oxidation, *ACS Appl. Nano Materials*, **2018**, *1*, 556-563.

533861 The Geode Process: A Route to the Large-Scale Manufacturing of Functionally-Encoded Nanostructures

Abstract:

A wide range of high-tech applications require constituent materials and devices that must simultaneously exhibit nanoscale control of compositional heterogeneity and be produced at large scales. Semiconductor nanowires, for example, can be produced with exquisite control of composition and morphology using the vapor-liquid-solid (VLS) mechanism that, however, remains limited to very small production rates. Here, we introduce the Geode process to synthesize functionally-encoded semiconductor nanowires at throughputs orders of magnitude beyond the state-of-the art. Central to the Geode process are sacrificial, porous-walled, seed particle-lined silica microcapsules, whose interior surface serves as a high-surface area growth substrate. Microcapsules protect the growing nanostructures, are produced with a scalable emulsion templating technique, and are compatible with large-scale chemical reactors. We will show how microcapsule structure and drying is influenced by silica nanoparticle type and concentration, emulsification parameters, and nanoparticle cross-linking agent. We will also demonstrate the synthesis of Si nanowires with programmable dopant profiles on the microcapsule interior. The impact of the microcapsule wall on the transport of reactive species will be characterized by examining changes to dopant profile.

537931 Invited Speaker: Carbon Nanotube-Based Optical Sensors for Cancer Detection

Abstract:

The early detection of cancer could lead to vastly improved patient outcomes. We aim to quantify cancer biomarkers within the body at early disease stages, permitting detection before symptoms arise. We are developing implantable nanosensors, using the unique optical properties of carbon nanotubes, to facilitate non-invasive detection through living tissues. The near-infrared emission of carbon nanotubes exhibits sensitivity to the local environment that can be controlled via non-covalent modification of the nanotube surface. We developed new methods to use carbon nanotube emission to transduce the binding of cancer biomarkers, including microRNA and proteins. We then developed implantable devices using porous membranes to allow diffusion of cancer biomarkers to reach the nanotubes. We found that the sensors could non-invasively detect the ovarian cancer biomarker in HE4 within orthotopic models of high grade serous ovarian carcinoma. The work portends clinical translation of implantable devices for use in patients with risk factors for disease to detect disease onset, recurrence, or to monitor treatment response.

534012 Peptide-DNA Hybrid Nanomaterials for Biology and Medicine

Abstract:

Peptides and DNA represent two of the most attractive categories of molecules for the construction of nanomaterials for biology and medicine. Peptides provide a rich palette of biological functionality and self-assembly behavior, and DNA can be used to construct complex nanostructures with programmable, dynamic properties. We sought to merge the advantages of these two molecular platforms through the use of peptide-DNA (P-DNA) hybrid materials. I will present recent progress in various areas: 1) Peptide-DNA nanotubes for guiding stem cell differentiation. We conjugated the cell adhesion peptide RGDS to single-stranded DNA, and used this hybrid molecule to construct DNA tile-based nanotubes displaying the peptide in a multivalent fashion. The nanotubes served as an efficient scaffold for neural stem cell adhesion and differentiation (NSCs). 2) Dynamic and programmable control of cell bioactivity: I will describe a unified molecular platform that can be programmed to control the dynamics, spatial positioning, and combinatorial synergies of signals in extracellular matrices. In this approach, a peptide-DNA (P-DNA) molecule is immobilized on a surface through DNA tethers. By engineering a series of tethers responsive to different stimuli, we show that cells adhered and spread on the surface reversibly. The use of P-DNA in cell signaling allowed multiple cycles of reversibility by simply adding soluble biologically compatible molecules such as DNA and enzymes. The DNA was also used as a molecular ruler to control the distance-dependent synergy between two adhesion peptides. Finally, orthogonal DNA tethers combined the binding of both a peptide and a bioactive protein to the surface in order to enhance cell differentiation. Multiple orthogonal DNA handles can be designed to allow for the selective presentation of different signals, with the ability to independently up- or down-regulate each over time. 3) Peptide amphiphile (PA)-DNA conjugates for programmable hydrogels. Peptide amphiphiles (PAs) consist of peptides bearing hydrophobic tails that allow their self-assembly into one-dimensional nanofibers ~10 nm in diameter and microns in length. These nanofibers can be used to construct highly potent bioactive hydrogels by incorporating functional epitopes. We synthesized PA-DNA molecules which, upon co-assembly with unmodified PAs, result in hybrid nanofibers that display multivalent DNA handles. The DNA can be used to reversibly crosslink the fibers into self-supporting gels, dynamically and reversibly present biological signals, or tune the mechanical properties of the resulting materials.

533981 Self-Assembled Hybrid Peptide-DNA and Protein-DNA Nanostructures

Abstract:

The ultimate goal of nanotechnology is to build structures on the 10-100 nm length scale, with control of matter down to the atomic level. In recent years, DNA has emerged as a powerful molecular building block for the construction of nanostructure materials due to the specificity of Watson-Crick pairing. However, DNA nanostructures are limited by the physical and chemical properties of oligonucleotides, which can be a hindrance for biological applications, or for attaining higher resolution placement of synthetic functionality. Proteins and peptides, by contrast, are the natural "language" of biology, possess a greater diversity of chemical groups, and can position materials sub-nanometer precision. In this work, we present the novel integration of two self-assembling polypeptide motifs with DNA nanostructures: 1) the coiled-coil, and 2) hetero-trimeric proteins. We have synthesized peptide- and protein-DNA conjugates with high site-specificity, allowing us to integrate these components with self-assembled DNA structures with high precision. DNA tiles, wireframe cages, and DNA origami were modified in specific locations with one of two heterodimeric coiled coil peptides that bind with sub-nanomolar affinity. The coiled coil interaction drove the hierarchical assembly of these components into dimers and one-dimensional nanofibers, providing an orthogonal assembly "code" to DNA hybridization. In addition, we will demonstrate the assembly of hierarchically-ordered 3D cages comprised of trimeric protein vertices with tunable DNA arms linking them. These materials are the first to our knowledge to integrate synthetic, self-assembling peptide/protein-DNA hybrids with well-established DNA motifs, and demonstrate the potential of protein-DNA nanotechnology.

537625 Dynamic Covalent Assembly of Abiotic, Information-Bearing Oligomers

Abstract:

Self-assembly processes are often based upon weak intermolecular interactions, one consequence of which is that the assembled structures are often fragile and susceptible to thermal or mechanical degradation. Covalent bonds typically exhibit bond energies that are over an order of magnitude higher than those for hydrogen bonds and could conceivably provide a route for the fabrication of far more robust assemblies; however, the selective creation of exquisite nanostructures by self-assembly and the toughness imparted by covalent bonds are generally perceived as mutually exclusive owing to the prevalent irreversibility of covalent bond-generating reactions. Fortunately, several covalent interactions are known to be reversible under particular reaction conditions, enabling the error correction mechanism that is essential for the selective fabrication of supramolecular structures. As a result of the enormously greater strength and directionality offered by covalent bonds in comparison to weak intermolecular interactions, these 'dynamic' covalent chemistries offer an elegant approach to nanostructure assembly that combines complexity AND toughness. Here, we will describe the synthesis and self-assembly of sequence specific oligomers mediated by dynamic covalent interactions, including the dynamic covalent assembly of molecular ladders with up to sixteen rungs between complementary peptoid oligomers bearing either amine, aldehyde, boronic acid, or catechol functional groups, the concurrent assembly of greater than two oligomeric precursor strands to yield long molecular ladders and wide molecular grids, and the *in situ* deprotection and subsequent, information-directed assembly of strands bearing both amine and aldehyde functional groups.

537578 Coffee-Ring Biomaterials As Nanoglues for Laser-Activated Tissue Sealing

Abstract:

Skin grafts are used to facilitate healing in cases of tissue loss from infections, deep burns, and large, open wounds that are too big to heal on their own. Several surgical techniques including sutures and tissue adhesives are used to restrain the graft onto the wound but can be traumatic, do not provide fluid tight seals, can leave wounds susceptible to infection, and lead to complications after surgery. There is a critical need to develop alternative surgical methods for more reliable skin grafting. Laser-activated tissue sealing provides an effective alternative approach in securing skin grafts to host tissue. In this approach, plasmonic nanoparticles, which convert incident laser light to heat, are embedded within biomaterials leading to the formation of photothermal nanocomposites. Incident near infrared laser light is converted to heat, which facilitates integration of the biomaterial with the tissue leading to efficient sealing. In the present study, we employed the coffee-ring effect to localize gold nanorods at the periphery of sealant biomaterials, including silk, chitosan and alginate, in order to facilitate sealing only at the edges and minimize thermal damage to tissue. The temperature rise following laser irradiation was determined experimentally and studied using mathematical modeling. The efficacy of these coffee-ring biomaterials was determined for sealing ex vivo tissues and in full-thickness wounds in mice skin. Recovery of tissue mechanical properties and histopathological analyses were used to determine the efficacy of sealing using laser activation of coffee-ring biomaterials as well as with conventional methods including suturing. Our studies indicate that coffee-ring biomaterials are promising laser-activated sealants in soft tissue approximation including in skin graft surgeries.

536184 Micropatterning of Silk Protein-Conductive Polymer Biocomposites for Fabrication of Flexible Devices

Abstract:

The precise micro- and nano fabrication of flexible devices using biomimetic materials has attracted significant attention owing to the potential for improved adaptability to physiological environments, as well as to form â€œgreenâ € devices for sustainable future. In particular, silk proteins (fibroin and sericin) are versatile materials to form flexible, biocompatible and biodegradable devices such as sensors, energy storage, and bio-microelectromechanical systems (bioMEMS). In conjunction with intrinsically conducting polymers (CPs) with their chemical diversity and ionic conductivity, the electrochemical functionalization of such biomaterials in organic bioelectronics can be enabled. In this talk, we will discuss various biodevices for sensing and energy storage that can be fabricated while retaining mechanical compliance, biocompatibility and degradability.

Biomimetic composites based on silk proteins and CPs are used to create novel photoreactive sensing inks, which are then coupled with lithographic tools to form micropatterned, organic, flexible and degradable biodevices. The biochemically modified silk proteins behave as negative-tone photoresists (fibroin and sericin photoresists) to enable accurate and high-resolution microfabrication on flexible substrates using photolithography. Optimization of the biofriendly fabrication process as well as the versatility of this biomaterial platform that can combine different CPs (e.g. PEDOT:PSS, polyaniline (PANI)) with protein microfabrication will be presented. The electrochemical characterization of the devices via electrochemical impedance spectroscopy (EIS) and other physical voltammetry techniques will be discussed towards the fabrication of microsupercapacitors and the detection of electroactive (e.g. dopamine and ascorbic acid) as well as electro-inactive (e.g. C-reactive protein) biomarkers. Importantly, the performance of the devices under mechanical flexure, and to form highly compliant, lightweight sensors that can integrate with soft interfaces will be shown. The engineered properties of these organic biocomposites provide opportunities as sustainable devices that can be programmed to resorb, adapt to the tissue or skin interface, and provide high resolution spatial and temporal sensing.

537857 Chitosan / Cellulose Nanocrystals / Calcium Phosphate Hydrogels for Vertebral Compression Fracture Treatment

Abstract:

Vertebral compression fractures are the most common consequence of osteoporosis accounting for approximately 700,000 out of the 1.5 million total osteoporotic fractures that occur annually in the USA. Injection of poly(methyl methacrylate), a non-biodegradable and non-bioactive polymer, into the vertebral bode as a mechanically stiff bone cement is the most common treatment for vertebral compression fractures. Recently there has been a growing interest to use bioactive, degradable, and mechanically suitable substitutes for the treatment of these fractures. In this research, we focused on creating injectable, chitosan-based hydrogels for the treatment of vertebra compression fractures. First, we investigated the effect of different ratios of ionic and covalent crosslinking on hydrogel physical stability and mechanical properties. Then, cellulose nanocrystals with varying surface charge were utilized to modify the mechanical properties and physical stability of the chitosan matrix. Finally, different calcium phosphates with various solubility were incorporated into the matrix in order to achieve sustained release of calcium and phosphate ions for optimal bone inductivity outcomes without inducing unwanted cytotoxicity.

525432 Young Investigator Award - Nanotechnology As a Tool to Study and Direct Immune Function

Abstract:

Our research combines immunology and biomaterials to i) understand the interactions between synthetic materials and immune tissues, and ii) to design more selective therapeutic vaccines for cancer and autoimmunity. This presentation will highlight our most recent efforts integrating nanotechnology, primary cell culture, animal models, and samples from human patients. One example will discuss our work using quantum dots as tools to study how the combinations, densities, and concentrations of self-antigens control the development of immune tolerance. Understanding these parameters could inform design of new therapies for disease such as multiple sclerosis and type 1 diabetes. A second example will highlight a new approach to control inflammatory pathways that are over-expressed during autoimmune disease ("toll-like receptors") by self-assembling immune signals into nanostructured materials. These materials are simpler than traditional polymeric materials because they are build entirely from immune signals, but mimic attractive features of biomaterials such as co-delivery and cargo protection. Because these materials can selectively reduce disease-driven inflammation during mouse models of multiple sclerosis, they could contribute to efficacious therapies that avoid the immunosuppression plaguing existing human autoimmune therapies.

540773 2018 Outlook for Energy: A View to 2040

Abstract:

The Outlook for Energy is ExxonMobil's global view of energy demand and supply through 2040. It also highlights the dual challenge of providing affordable energy to support prosperity while reducing environmental impacts – including the risks of climate change. We use the Outlook to help inform our long-term business strategies and investment plans.
The Outlook anticipates significant changes through 2040 across the world to boost living standards, reshape the use of energy, broaden access to abundant energy supplies, and accelerate decarbonization of the world's energy system to address the risks of climate change.

- Global energy needs will rise about 25 percent over the period to 2040
- Electricity demand will nearly double in non-OECD countries
- Worldwide electricity from solar and wind will increase about 400 percent
- Natural gas will expand its role, led by growth for electricity generation
- Growth in oil demand will be driven by commercial transportation and chemical needs
- Efficiency gains and growing use of less-carbon intensive energy sources will contribute to a nearly 45 percent decline in the carbon intensity of global GDP

Energy is fundamental to modern life, and as the world's population approaches 9 billion people in 2040, we are challenged to improve living standards everywhere. Understanding the factors that drive the world's energy needs – and likely choices to meet those needs – is the mission of the Outlook. By sharing the Outlook with the public, we hope to broaden that understanding. Energy matters to everyone, and we all play a role in shaping its future.

541543 Energy Decarbonisation Scenarios

Abstract:

With the energy sector being associated with 2/3 of the annual greenhouse gas emissions, decarbonizing the energy is a key step to reduce their level. The National Determined Contributions (NDC's) announced so far by the participating countries do not meet the objectives of the Paris-climate agreement, as far as reaching a peak in emissions in the near future or decreasing to a level that is compatible with the stated long-term targets.

Using different scenarios for energy demand and supply, options for energy sector decarbonisation are evaluated. Meeting the 2-degree target requires a major transformation in the power sector, efficiency improvements – specially in the buildings and in the industry, and a revolution in transport towards fuel and modal shifts. With a significant increase in renewable energy, system integration and optimization will also be key in reducing the cost of the energy transition. Meeting more ambitious climate goals ("well below 2 degrees") will push further the technological innovation and implementation requirements throughout the entire energy value chain. A scenario aiming at zero net-emissions by 2060 is presented, along with a comparison to the 2-degree case.

542625 Fundamental Research Needs to Advance Energy Technologies

Abstract:

The U. S. Department of Energy (DOE), Office of Science, Office of Basic Energy Sciences (BES) supports fundamental research in chemical and materials sciences to provide the foundations for new energy technologies and to support DOE missions in energy, environment, and national security. This presentation will discuss opportunities for fundamental research to impact DOE's energy mission "to catalyze the timely, material, and efficient transformation of the nation's energy system and secure U.S. leadership in energy technologies" with a focus on the way we generate, store and use energy nationally. I will provide an overview of BES strategic planning over the past decade that identified priority research directions for advancing energy applications, highlight key scientific advances in these areas, and discuss some future opportunities for modern science, particularly chemical sciences, to accelerate the transformation of the U. S. energy portfolio.

527113 Application of Novel Porous Graphene Nanoplatelets Composites for Enhanced Heat Transfer Properties

Abstract:

Pool boiling or two phase heat transfer techniques dissipate a large amount of heat by maintaining low surface temperatures as compared to single-liquid phase based heat transfer processes. The maximum amount of heat that can be dissipated from the surface is determined by critical heat flux (CHF), while the efficiency of the boiling is determined by heat transfer coefficient (HTC). Pool boiling finds applications where the devices generate a large amount of heat, such as high-power density electronic devices, nuclear reactors, reboilers, and heat exchangers. The pool boiling performance can further be improved by surface enhancement techniques such as micro/nano porous coatings and by implementing high thermal conductivity materials. In the recent years several carbon-based nanomaterials have been exploited for pool boiling applications due to their high thermal properties and surface areas

We investigated carbon nanotubes, multilayered chemical vapor deposited graphene/graphene oxide, and graphene nanoplatelets (GNP) coatings and achieved higher pool boiling performance with graphene nanoplatelets (GNP), however, the GNP coatings did not sustain the repetitive pool boiling tests making them ineffective for industrial applications. To overcome this, we developed novel copper-GNP (Cu-GNP) porous coatings by a multistep electrodeposition technique owing to higher thermal properties of copper particles.

Hierarchal porous surfaces with diverse morphologies were obtained by varying the electrodeposition process parameters and the concentration of GNP in the composites which were confirmed by scanning electron microscope (SEM). The rendered surfaces were hydrophilic in nature possessing higher wickability. The durability of the coatings was confirmed using the repetitive pool boiling studies. The composite coating with 2% GNP (weight/volume) produced a very high CHF of 286 W/cm^2 and HTC of 204 kW/m^2-°C, representing an improvement of 130% in CHF and 290% in HTC compared to pristine copper surface. The proposed mechanism of enhanced heat transfer properties is attributed to the surface pores that serve as nucleation sites for additional bubble formation that depart in upward direction creating a turbulent convection that further improves the heat transfer rate.

533903 1-Dimensional Carbon Nanoparticles for Functional Biomolecule Delivery to Mature Plants

Abstract:

Plants are at the core of environmental sustainability efforts, natural product synthesis of pharmaceuticals, and food security for a growing population under a changing global climate. Traditional plant breeding strategies are slow, laborious, and lack precise control over plant genotype and phenotype. Over the past several decades, remarkable progress has been made in biotechnology with the improvement of genome editing and sequencing tools. Owing to these recent advancements, plant bioengineering can be leveraged to create crops that are resistant to disease and drought. In pharmaceuticals and therapeutics, genetically engineered plants can be used to synthesize valuable small-molecule drugs and recombinant proteins. Furthermore, bioengineered plants may provide cleaner and more efficient biofuels.

Despite the promise of plant biotechnology, the presence of the plant cell wall has posed a transport limitation to delivery of exogenous biomolecules to plant cells, such that conventional methods of biomolecule delivery (*Agrobacterium* and gene gun) are inefficacious for plant biotechnology[1]. Thus, the plant cell wall is the primary bottleneck of plant genetic engineering. To date, plant biotechnology lacks a method that allows passive delivery of diverse biomolecules into a broad range of plant species without the aid of external force and without causing tissue damage. We posit **carbon nanotechnology** as a key driver in the creation of a transformational tool to address delivery challenges in plants, and to enhance the throughput of plant genetic engineering.

Under certain surface chemistries, high aspect ratio nanomaterials such as carbon nanotubes (CNTs) have shown to traverse extracted chloroplast and plant membranes with several figures of merit: high aspect ratio, exceptional tensile strength, high surface area-to-volume ratio, and biocompatibility. When bound to CNTs, biomolecules are protected from cellular degradation, exhibiting superior biostability compared to free biomolecules. Herein, we present a CNT-based platform that permits diverse conjugation chemistries to deliver functional biomolecules into both model and consumer food crop plants with high efficiency and no toxicity[2]. Surface-modified CNTs are covalently-functionalized with cationic polymers (such as polyethylenimine - PEI) to load DNA via electrostatic attraction, for DNA delivery into mature arugula (dicot) and wheat (monocot) leaves. Through this platform, we obtain strong transient protein expression, with efficiencies comparable to *Agrobacterium*-mediated and higher than gene gun gene delivery. We also show nanotube-based transient protein expression in arugula protoplasts with 85% transformation efficiency. Lastly, we achieve 95% transient gene silencing in *Nicotiana benthamiana* leaves through the delivery of small interfering RNA molecules with pristine SWCNTs. This study establishes efficient transient gene expression and silencing in mature plants through passive carbon nanotube-mediated delivery of functional biomolecules, and can enable high-throughput genetic plant transformations for a variety of plant biotechnology applications.

1. Demirer, G.S., Landry, M.P. Delivering Genes to Plants. AIChE SBE (2017)
2. Demirer, G.S. *et al.* (2018) High Aspect Ratio Nanomaterials Enable Biomolecule Delivery and Transgene Expression or Silencing in Mature Plants. *bioRxiv* DOI: 10.1101/179549

527285 Electricity from Asymmetric Chemical Doping of Single-Walled Carbon Nanotubes

Abstract:

There is a pressing need to find alternatives to conventional energy generation techniques, specifically those that power micro- and nanoelectronic systems with upper bounds on device sizes. The concept of electricity generation based on asymmetric chemical doping (ACD) of single-walled carbon nanotubes is presented.

Specifically, ACD enables unique power sources driven entirely by a chemical potential gradient of the electrical carriers within the nanomaterial, established through a spatially asymmetric interaction with the surrounding chemical environment. Short circuit electrical current (11.9 μA mg^{-1}) and open circuit potential, in excess of 1.0 V, can be reversibly produced by localized acetonitrile (CH_3CN) doping under ambient conditions. An inverse length-scaling of the maximum power as $L^{-1.03}$ that creates specific powers as large as 30.0 kW kg^{-1} highlights ACD's potential for microscale energy generation.

In this talk, I will introduce the electron transfer mechanism that describes the electricity generation process in ACD, as well as our efforts to evolve from organic-based molecular dopants towards a system powered entirely through interactions with water molecules. Apart from being used as a power source for environment-interfacing microelectronic systems, which I will discuss on the basis of a Colloidal State Machine invented recently, ACD also finds interesting applications in electrochemistry. In particular, it allows us to construct a particulate platform that generates "packets" of electricity on demand, in solution, and drives electrochemical transformations in situ by virtue of interacting with the solvent.

References:

(1) Liu, A. T.;* Kunai, Y.;* Liu, P.; Kaplan, A.; Cottrill, A. L.; Smith-Dell, J. S.; Strano, M. S. *Adv. Mater.* **2016**, *28*, 9752.

(2) Liu, A. T.;* Mahajan, S. G.;* Cottrill, A. L.; Kunai, Y.; Bender, D.; Castillo, J.; Gibbs, S. L.; Strano, M. S. *Energy & Environmental Science* **2016**, *9*, 1290

(3) Liu, A. T.;* Kunai, Y.;* Cottrill, A. L.; Koman, V. B.; Liu, P.; Kozawa, D.; Gong, X.; Strano, M. S. *J. Am. Chem. Soc.* **2017**, *139*, 15328.

531068 Graphene Oxide Nanocomposite Hydrogels Capable of Wastewater Dye Sequestration

Abstract:

Organic dyes are widely used in many chemical industries and, unfortunately, they are often discharged with wastewater without adequate treatment, becoming one of the main causes of harmful water pollution. Methylene blue (MB) is a cationic dye, and like many other cationic dyes, it is made from known carcinogenic aromatic compounds that can cause severe environmental problems and health risks. Various processes have been developed to purify water, such as membrane filtration, ion exchange, chemical oxidation, etc. However, they have many restrictions such as lack of practicality, formation of toxic by-products, production of high amounts of residues, or incapability of regeneration. Among these methods, adsorption has been preferred and widely used lately because of its comparable low cost, simplicity of operation, low generation of residues and recycling capabilities of the adsorbent. In recent years, graphene oxide (GO) has attracted attention as a potential dye adsorbent because of its unique two-dimensional (2D) structure with oxygen functional groups, a large surface area, strong Ï€-Ï€ interactions and highly negatively-charged surface, allowing a superior adsorption of aromatic cationic dyes. Hydrogels are networks of crosslinked hydrophilic polymer and can potentially serve as three-dimensional templates to immobilize and present GO in the aqueous environments demanded by wastewater treatment applications. Thus, GO-laden hydrogels have the potential to efficiently sequester cationic dyes from wastewater and retain them for transport and disposal.

In this study, GO was integrated into hydrogels by entrapping GO in networks of polymerized acrylamide (AAm) crosslinked with polyethylene glycol dimethacrylate (PEGDMA). Both nanoporous gel structures with and without GO were investigated as candidate wastewater treatment materials. The porosity and swelling ratio of the gels were determined, and morphological characterizations of gels were carried out using scanning electron microscopy (SEM). MB and eosin Y (EY) were chosen as model adsorbates to evaluate and compare the adsorption capabilities of GO gels for cationic and anionic dyes, respectively. Samples from GO gel adsorption kinetic studies were analyzed via UV-vis spectrometry in order to assess dye sequestration capabilities. The isotherms at different dye concentrations, kinetic studies and regeneration of the gels are currently ongoing. So far, GO-laden and GO-free hydrogels showed similar inabilities to sequester anionic EY removal, as expected. However, GO-laden hydrogels exhibited the ability to efficiently remove cationic MB. Finally, dye-adsorbed GO hydrogels can easily be removed from the aqueous environment, suggesting a potential alternative for organic cationic dye removal in wastewater treatment.

536865 Addressing the Isomer Cataloging Problem for Nanopores in Graphene and Other 2D Materials

Abstract:

Extended defects or nanopores in graphene and other 2D materials, such as hexagonal boron nitride (hBN), can be used to tailor their electronic, magnetic, electrochemical, and barrier membrane properties. However, the existence of a large number of possible lattice isomers of nanopores makes their quantitative study a seemingly intractable problem, while confounding the interpretation of experimental and simulated data. Herein, we formulate an *ab initio* solution to this *Isomer Cataloging Problem (ICP)*, which combines extensive electronic-structure calculations, kinetic Monte Carlo simulations, and chemical graph theory, consistent with the underlying symmetries of 2D materials, to generate a catalog of unique, most-probable isomers of 2D lattice nanopores. The calculated first-principles data set provides precise, experimentally consistent rates for transitions between nanopore shapes in the graphene lattice. Further, the results demonstrate remarkable agreement of our model with precise nanopore shapes observed in graphene using transmission electron microscopy, and show that the thermodynamic stability of a nanopore in graphene is distinct from its kinetic stability. The methodology also predicts the experimentally observed prevalence of triangular nanopores in hBN, supporting the assertion that it can solve a wide range of ICPs for other 2D membranes and lattices, thereby providing a much-needed connection between molecular design and fabrication. The methodology developed herein should accelerate the application of nanoporous graphene and other 2D materials for optoelectronics, magnetism, gas separations, desalination, and biological applications, by establishing specific links between experiment and theory/simulations through the solutions of ICPs.

516930 Twenty Years of Metal Nanoparticle Synthesis Using Biotemplates

Abstract:

Biomolecules have surface functionalities that can promote inorganic mineralization at very mild reaction conditions and in non-toxic aqueous solutions. Tobacco mosaic virus (TMV) and barley striped mosaic virus (BSMV) biotemplates are used to control the shape of the final nanostructures and also the size of the primary metal particles. This presentation will discuss the evolution of research focused on the synthesis of metal coatings on these biotemplates.

527599 Theoretical and Experimental Study of Germanium Nanoparticle Formation in a Controlled Nucleation Gas Phase Process

Abstract:

Semiconducting nanoparticles out of germanium are promising materials for the application in printable electronics. Dispersed in organic liquids and deposited as thin functional film, e.g. as a thin film transistor (Meric et al., 2015). The synthesis of these nanoparticles (NPs) is challenging because the homogenous pyrolysis of the precursor leads to high nucleation rates with subsequent agglomeration and sintering. One strategy to overcome this problem is the introduction of seed particles to suppress the homogenous nucleation (Alam and Flagan, 1986). Therefore, we introduced an external seeding unit supporting the synthesis of semiconducting nanoparticles in a hot-wall reactor. This external unit allows a detailed investigation of the effect of seed particles on the particle growth. The experimental observations are supported theoretically by population balance modelling.

For the experiments, a hot-wall reactor (HWR) is combined with a hot-wire generator (HWG). The HWG is used to produce small metallic seed particles by condensation from a supersaturated vapor. This vapor is generated by passing an electric current through a metallic wire held between to electrodes. The seed particle stream is directly mixed with the precursor (GeH_4) and injected into the hot-wall reactor. In this stage, the NPs grow on the seed particles at temperatures between 700 °C and 1000 °C. At the reactor exit the aerosol is quenched by nitrogen. The particles are sampled *via* a low-pressure impactor by directly deposition on a substrate or as powder *via* a filter. For the theoretical investigation a one-dimensional model, which was already successfully introduced in previous works for the formation of silicon nanoparticles (Körmer at al., 2010), was transferred to the germanium system. The influence of the seed and germane concentration and temperature is studied.

The aerosol properties of the seed particles have a significant influence on the Ge NP morphology and can be controlled for example by the used metal and the applied voltage. The seed concentration can be varied in the range from 10^5 - 10^8 1/cm^3 and the size from 3 - 15 nm. An increasing voltage leads to a higher wire temperature and this consequently to a higher amount of evaporated material. The seed particle concentration in the reactor increases and homogeneous nucleation is suppressed at a certain threshold (~ 10^{-6} - 10 ^{7} 1/cm^3).

The final seeded particles show a well-defined morphology with a size of around 35 nm (Geometric standard deviation GSD: 1.08). XRD measurements show the typical pattern of Bragg peaks for crystalline germanium. A quantitative Rietveld analysis gave a crystallite size of around 31 nm. This value is in good accordance to PSD evaluation by SEM images. The discrepancy may be due to the amorphous oxide layer, which is 3 to 4 nm thick. With increasing partial pressure, the particle grows larger in size, while the morphology and a narrow size distribution are preserved. Even a six-fold increase in the germane partial pressure only leads to a slightly broadening of the particle size distribution (GSD: 1.12). Therefore, narrow

distributed Ge NPs can be produced supported by the HWG with a particle size in the range from 25 to 50 nm. The synthesis temperature has no significant influence on the particle properties. These observations can also be seen in the theoretical results. The model shows that the key parameter is the seed concentration, which has a strong influence on the homogenous nucleation rate and the final particle size. The temperature has no impact on particle formation because already at 700 °C a complete conversion of germane takes place in the presence of seed particles. The predicted particle distributions are also in good accordance to the experimental results.

The progress on the synthesis of Ge NPs supported by external seeding will be demonstrated. The influence of the seeding unit and the process parameters on particle formation and properties will be discussed experimentally and theoretical in detail.

This work was supported by the German Research Council (DFG) and the Cluster of Excellence "Engineering of Advanced Materials" (EAM).

Meric, Z., Mehringer, C., Karpstein, N., Jank, M.P.M., Peukert, W., and Frey, L. (2015). Tunable conduction type of solution-processed germanium nanoparticle based field effect transistors and their inverter integration. Phys. Chem. Chem. Phys., 17(34):22106–22114.

Alam, M.K. and Flagan, R.C. (1986). Controlled nucleation aerosol reactors: production of bulk silicon. Aerosol Sci. Technol., 5(2):237–248.

Körmer, R., Schmid, H.J., and Peukert, W. (2010). Aerosol synthesis of silicon nanoparticles with narrow size distribution-Part 2: Theoretical analysis of the formation mechanism. J. Aerosol Sci., 41(11):1008–1019.

535403 Tuning the Optical, Catalytic, and Physical Properties of CuO Nanosheets Using Organic Functional Groups

Abstract:

A new class of hydrophobic CuO nanosheets is introduced by functionalization of the cupric oxide surface with p-xylene, toluene, hexane, methylcyclohexane and chlorobenzene. The resulting nanosheets exhibit a wide range of hydrophobicities with contact angles ranging from 146° (p-xylene) to 27° (chlorobenzene). Significant changes to the surface composition induced by functionalization was verified by XPS, ATR-FTIR and computational modeling. UV-Vis DRS showed that functionalization with these molecules increase the band gap of the CuO nanosheets by 0.05-0.08eV. DFT and tight-binding calculations showed near identical increases of band gaps in modelled slabs of CuO with functional groups attached. Furthermore, TGA and H_2-TPR show that the functional groups stabilize the CuO nanosheets, and that the stability is correlated with the hydrophobicity of the material.

534856 Spatial Atomic Layer Deposition By "Air Hockey" Design for Dielectric Multilayer Optical Films

Abstract:

Spatial Atomic Layer Deposition by "Air Hockey" Design for Dielectric Multilayer Optical Films

J. Grasso, N. Oliveira, B. Willis

University of Connecticut

Dielectric multilayer structures exhibit exceptional specular properties for employment in various optical applications; such as, filters, antireflection and high reflection coatings. Specular properties can be tailored for specific applications by altering the composition of the materials in the stack. Atomic layer deposition (ALD) is one proven method used for constructing multilayer stacks. The conformity, atomic-level thickness control, and freedom to choose countless materials associated with films deposited by ALD are advantageous to the performance of sophisticated multilayer optical devices. Dielectric multilayer optical films can prevent radiation lost from objects at high temperatures by altering the optical properties of the cold-side; leading to the resurrection of the incandescent source. These complex films are considerably thick and are unreasonable to fabricate with current technology. Slow deposition rates for conventional ALD contribute to the impracticality of utilization of this technique for high throughput processing. Previous work reports that 20 layer stacks take upwards of 16 hours to complete. Spatial ALD operates by continuously supplying the precursors while keeping them physically separated in different zones; eliminating the essential time intensive purging step in conventional ALD. The deposition rate is drastically improved in spatial ALD due to the enhancement of the mass transfer of precursors to the substrate, transforming the process to being kinetically limited. The magnitude of this effect may be correlated with the distance separating the precursor inlets and the substrate. Advances in the deposition rate improvement requires an extensive understanding of the fluid dynamics of the gases and its dependence of the process parameters.

We present a case study for the well understood ALD of alumina by trimethylaluminum (TMA) and water using a novel spatial ALD system operating analogous to an air hockey table. Dispersed nitrogen inlets float the substrate allowing for virtually frictionless motion amplifying its linear velocity. Discrete introduction of the precursors is achieved via a nitrogen barrier which prevents undesirable mixing of the precursors as the substrate travels across the injector region. Each injector vent is surrounded by two vacuum pumps to eliminate excess gas not utilized in the reaction. Exploratory experiments have been conducted in attempt to understand the affect that precursor ratios and gap height have on performance. Spectroscopic ellipsometry measurements are used to study the optical properties and measure the thickness of the thin epitaxial films. 1D laser displacement measurements illustrate the ability to control the gap height by adjusting the flow rate of the table gas; height control within several microns can be achieved. A full factorial design of experiments is being implemented to determine the key factors that contribute to film uniformity.

The fluid dynamics under the substrate are not well understood and computational efforts are being made in COMSOL Multiphysics to gain insight to the underlying physics. It is believed that the gap height may be the most significant factor in determining the quality of film deposited. In comparison to the size of substrate, the gap height is significantly smaller in magnitude and therefore the underlying fluid dynamics can be determined by the thin gap approximation, or lubrication theory. Lubrication theory can be used to correlate the pressure distribution under the substrate with the gap height. The gap height has a substantial influence on the behavior of the fluid motion. The fluid flow between an inlet and exhaust vent under the moving substrate can be simulated as Couette flow with a pressure gradient. When the desired direction of flow is opposite the motion of the substrate an adverse pressure gradient is present and backflow can occur. This phenomenon is strongly dependent on the gap height and the vent spacing; it was found to occur when the gap height is small and the vent spacing is large. If operating the system in the backflow regime, undesired mixing of the precursors would result and chemical vapor deposition behavior would be observed. This places a constraint on the system to ensure this operation does not occur. The effectiveness of the nitrogen curtain as a diffusional barrier can also be analyzed in COMSOL and will affect the design of the system and operating conditions. Further understanding of the system will allow for optimization of the design.

533543 Direct Biomineralizalization and Integration of Heterostructured Nanomaterials into Quantum Dot Sensitized Solar Cells

Abstract:

Functional nanomaterials are at the core of many innovations in renewable energy generation and catalysis. However, many of these materials are currently synthesized through high temperature, multi-step processes that utilize organic solvents and expensive precursors. These factors increase the complexity and economic and environmental costs of manufacturing scale-up. In contrast, biomineralization, the process by which biological systems produce structural nanomaterials, occurs under ambient conditions in aqueous media utilizing the available precursors. We have developed routes to adapt these biological processes to the direct and scalable, aqueous phase, controlled synthesis of a variety of functional nanoparticles; from core-shell metal chalcogenide quantum dots for solar harvesting to mixed metal oxide catalysts. Previous bioinspired synthesis routes typically utilize a biomolecule for templating of a chemical mineralization step, or lack tight control over size or composition of the particles. In contrast, our single enzyme approaches both template growth and catalyze mineralization to control size, composition and structure of the particles with function comparable to those synthesized by traditional chemical routes. In this talk, we will discuss this biomineralization process and routes towards integration of these particles in quantum dot sensitized solar cells.

527419 Monitoring Seed Formation Dynamics of Bulk-Nucleated Vapor-Solid-Solid Germanium Nanowires Via Resistance Measurements

Abstract:

Nanowire growth from bulk metal surfaces presents several scientifically interesting and technologically important research challenges. Scientifically, many knowledge gaps remain about the fundamental mechanism by which nanowires nucleate from bulk surfaces. Technologically, this approach presents a promising pathway towards large-scale continuous nanowire fabrication. We present a novel method to probe nanowire growth in-situ by monitoring the resistance of the substrate from which the nanowires nucleate. We show that probing the resistance of the inductively heated thin film provides real time diagnostic information about the reaction, including temperature, surface composition, and coarsening rate. We relate distinct features in the transient resistance to growth processes using *ex-situ* characterization via X-ray diffraction, scanning electron microscopy, energy-dispersive X-ray spectroscopy, optical microscopy, and contact profilometry. Using this technique, we reveal temporal intricacies of seed formation and supersaturation. We correlate the evolution of the germanide grain size to the nanowire diameter and provide guidance for diameter-specific nanowire synthesis. We extend the use of this technique to commercially scalable substrate materials. This method can enable rapid tuning of kinetic conditions of commercially scaled nanowire growth systems.

536752 Quality By Design in Nanomedicine: Application to a Microemulsion Delivery System

Abstract:

Nanomedicine is a concept which makes use of nano-scale carriers to provide therapeutic or diagnostic benefits in medicine. Despite promise of increased therapeutic efficacy, uncertainty of safety and translatability of nanomedicine platforms remain prevalent. Quality by design (QbD) is a practice used to heighten process safety and assure product quality. The methodology eases the development of pharmaceutical products by deliberate process that can lead to the procurement of high quality, effective nanoformulations while minimizing risk, time, and resources. QbD builds quality into a product by first establishing the overall objectives and defining ways in which to control the process. Although its use is strongly encouraged in the pharmaceutical industry, it is underutilized in the rapidly expanding field of nanomedicine. The presented work exemplifies a case study in which a QbD strategy is used in robust formulation development and highlights the benefits accrued as a result of applying QbD to nanomedicine, employing microemulsions as a model nanoformulation.

Microemulsions, a type of nanomedicine characterized by droplet diameter of 15-50 nm, are thermodynamically stable colloidal dispersions which are used to increase drug solubility and half-life. Microemulsions are safe, given that they can be developed with biodegradable, generally regarded as safe (GRAS) oils and surfactants. For these reasons, we chose microemulsions as our model nanoformulation. First, the quality target product profile (QTPP) of the microemulsion product was defined. After defining the associated critical quality attributes (CQAs), critical material attributes (CMAs), and critical process parameters (CPPs), we compiled a library of modes of failure during product manufacturing and ranked them using failure modes, effects, and criticality analysis (FMECA). We used the results of risk assessment to create a design of experiments (DOE) to which we fit multiple linear regression and logistic regression models to predict microemulsion properties and probability of meeting CQA specifications. We performed comprehensive quality control analyses to evaluate CQAs and overall were able to identify stable microemulsion formulations in a timely manner using a small number of experimental runs. Finally, we demonstrate the loading capability with multiple BCS class II drugs to demonstrate the versatility of this formulation and investigate the scale-up of the optimized microemulsion. Adapting this QbD methodology to other nanomedicine products is of value for time and resource minimization as well as smoothing the path to regulatory approval.

540091 Polymer Coated Gold-Ferric Oxide Superparamagnetic Nanoparticles for Theranostic Applications

Abstract:

A multifunctional magnetic nanoparticle (NP) is designed to perform a near-infrared (NIR)-responsive remote control photothermal ablation for the treatment of breast cancer. In contrast to the previously reported studies of gold (Au) magnetic (Fe3O4) core-shell nanoparticles (NPs), a Janus-like nanostructure is synthesized with Fe3O4 NPs decorated with Au resulting in an approximate size of 60 nm mean diameter. Epigallocatechin-3-gallate (EGCG), a ubiquitous bioactive polyphenol available in green tea, was adsorbed on the surface of Au–Fe3O4 NPs as a cell membrane targeting and cell growth inhibition ligand. Poly- l-lysine (PLL) was used for the surface functionalization of trisoctahedral Au–Fe3O4 NPs to deliver the NPs inside cells and function as a hydrophilic block for steric stabilization of the NPs in biological media. The PLL coating increased the colloidal stability and robustness of Au–Fe3O4 NPs (PLL–Au–Fe3O4). The unique properties of both Au surface plasmon resonance and superparamagnetic moment result in a multimodal platform for use as a nanothermal ablator and also as a magnetic resonance imaging (MRI) contrast agent, respectively. Taking advantage of the photothermal therapy, PLL–Au–Fe3O4 NPs were incubated with BT-474 and MDA-MB-231 breast cancer cells, investigated for the cytotoxicity and intracellular uptake, and remotely triggered by a NIR laser of ~808 nm (1 W/cm2 for 10 min). It is revealed that no significant (<10%) cytotoxicity was induced by PLL–Au–Fe3O4 NPs itself in BT-474 and MDA-MB-231 cells at concentrations up to 100 mg/ml. As soon as the NPs were triggered by a photothermal NIR laser, approximately 40 and 60% of BT-474 and MDA-MB-231 breast cancer cell death, respectively were observed using 100 mg/ml PLL–Au–Fe3O4 NPs, while the cells with the NPs but without laser showed <10% cell death compared to no laser treatment control. Combined together, the results demonstrate a new polymer gold superparamagnetic nanostructure that integrates both diagnostics function andphotothermal ablation of tumors into a single multimodal nanoplatform exhibiting a significant cancer cell death.

538395 Development of Steroid Biosensors Using Corona Phase Molecular Recognition and Translation to Physiological Biologging

Abstract:

For decades, biologgers have been tagging animals with various types of sensors to study aspects of their behavior unable to be queries in controlled laboratory environments. To date, environmental and physiological chemical information remains an underutilized data source that offers tremendous potential to raise the level of insight gained from such studies. In particular, continuous measurements of *in vivo* steroid levels may elucidate numerous physiological processes underlying animal behavior, yielding more sophisticated insights previously impossible. In this work, we describe our progress in utilizing a template steroid strategy coupled with Corona Phase Molecular Recognition (CoPhMoRe) to detect cortisol and progesterone. Polymers incorporating a template steroid in the backbone were used to suspend single walled carbon nanotubes (SWNT), creating a steroid library that was then screened against an analyte library of physiologically significant steroid hormones. Constructs selective for progesterone and cortisol will be presented. Our current efforts to translate CoPhMoRe sensors for marine organism physiological biologging will also be presented. We show the feasibility of using inexpensive cameras to detect changes in analyte concentrations, as well as demonstrate fluorescence measurements in fish tissue samples.

538759 Self-Assembly of Graphene/Noble Metal Nanotube Composite Electrodes for Fuel Cells and Supercapacitors

Abstract:

Self-Assembly of Graphene/Noble Metal Nanotube Composite Electrodes for Fuel Cells and Supercapacitors

Gabrielle M. Milanesa, Alexander N. Mitropoulos, Kamil Woronowicz, F. John Burpo, Enoch A. Nagelli*

Department of Chemistry & Life Science

United States Military Academy, West Point, New York 10996

*Corresponding PI: Dr. Enoch Nagelli, Email: enoch.nagelli@usma.edu

Graphene is a promising nanoscale platform to construct next generation three-dimensional electrodes for high performance, light-weight, and configurable energy storage devices. However, the design, synthesis, and assembly of nanomaterials with well-defined structure and the connection from individual nanoscale entities to these real world applications continues to present big challenges. This is critical to build nanostructured electrodes to develop high performance batteries and fuel cells. In order to achieve this, we demonstrate a scalable method using a solution based electrostatic self-assembly technique for the formation of novel 3D noble metal micro- or nanotube composites with graphene. For example, we electrostatically bind graphene oxide with platinum salt coordination complexes and chemically reduce the resultant 3D structure using sodium borohydride. Our design process enables the controlled formation and assembly of micro- or nanotubes onto 2D graphene sheets in order to develop new high surface area composite electrodes with any pure noble metal or alloy noble metals. This novel method of aqueous assembly is a platform which can lead to research and development of nanocomposite materials and structures that will enable development of high performance batteries and fuel cell electrodes.

References:

1. Georgakilas, M. Otyepka, A. B. Bourlinos,V. Chandra, N. Kim, K. C. Kemp, P. Hobza, R. Zboril, K. S. Kim, "Functionalization of Graphene: Covalent and Non-Covalent Approaches, Derivatives and Applications" *Chem. Rev.* **2012**, *112*, 61564.

2. Burpo, E. A. Nagelli, L. Morris, J. McClure, M. Ryu, J. Palmer, â€œDirect solution-based reduction synthesis of Au, Pd, and Pt aerogelsâ€ *J. Mater. Res.* **2017**, 22, 4153.

538803 Adipose Tissue Stem Cells Bioengineered in Nano-Biomimetic Col Scaffolds for Skin Tissue Engineering

Abstract:

Background: There is a motivation for tissue engineering, which is defined as promotion of repair or regeneration of injured or malfunctioned tissue or organ. In this way, tissue engineering (TE) is employed for helping the patient body and to produce materials which mimic the bodyâ€s in vivo tissue conditions. In addition, TE strategies showed promising results for improving the quality of regenerative therapeutics. Three dimensional (3D) biodegradable and biomimetic porous scaffolds are ideal frame works for skin tissue engineering.

Methods: In this study hybrid constructs of 3D scaffolds were fabricated by freeze-drying method from combinations of the type I Col, and synthetic poly lactic acid (PLLA) or poly capro lactone (PCL). Four groups of 3D porous scaffolds were fabricated and characterized by HNMR, FT-IR, SEM including: PCL, PCL-Col, PCL-PLLA, PCL-PLLA-Col. Adipose tissue derived mesenchymal stem cells (AT-MSCs) were seeded in those scaffolds and the viability, proliferation and adhesion of them were investigated using MTT assay and scanning electron microscopy (SEM).

Results and Discussion: The results showed that scaffolds containing Col in particular PCL-PLLA-Col with pore sizes near to 400nm and sufficiently interconnected , have significant greater potential ($p<0.01$) for encouraging AT-MSCs adhesion and growth.

Conclusion: The PCL-PLLA provided a mechanically strong mesh support and the type I Col microsponges encouraged cell adhesion and tissue formation. The outcomes of this study could be applicable for preparing artificial skin constructs.

538823 Self-Assembly of 3D Graphene/Carbon Nanotube Electrodes Via Poly(acrylic) Acid/Nickel Complexing for Biosensor Applications

Abstract:

Self-Assembly of 3D Graphene/Carbon Nanotube Electrodes via Poly(acrylic) Acid/Nickel Complexing for Biosensor Applications
An B. Vu, Kamil Woronowicz, Alexander N. Mitropoulos, F. John Burpo, Enoch A. Nagelli*
Department of Chemistry & Life Science
United States Military Academy, West Point, New York 10996
*Corresponding PI: Dr. Enoch Nagelli, Email: enoch.nagelli@usma.edu
The design, synthesis, and assembly of 2D graphene sheets with 1D carbon nanotubes (CNTs) can result in 3D nanostructures which can lead to the development of new biological/chemical sensors. Therefore, we developed a scalable aqueous solution-based process using poly(acrylic) acid (PAA)/CNT/graphene composite for advanced ultra-lightweight nanocomposites that are mechanically configurable and highly conductive. Using electrostatic self-assembly, we demonstrate an easy and scalable way to bind PAA/CNT composites with graphene oxide-coordinated Ni2+. The negative charge of many of the side chains of PAA in neutral pH water electrostatically bind with the Ni2+/graphene oxide to result in a polyelectrolyte-stabilized 3D carbon nanomaterial with a Ni2+ junction. This resulting 3D nanostructure is chemically reduced with sodium borohydride to form an all conductive polyelectrolyte/CNT-Ni nanoparticle â€“graphene composite. Therefore, our efforts are to contribute to the fundamental understanding of how specific carbon nanomaterials and metal nanoparticles and the combination of both can alter electronic band structure, doping, and affinity for other nanomaterials to be able to detect chemical and biological agents/analytes as well as transport hole or electrons through the 3D nanostructure.

538864 3D Carbon Nanomaterial/Platinum Microtube Composites for Oxygen Reduction Reaction Electrocatalysis in Fuel Cells

Abstract:

3D Carbon Nanomaterial/Platinum Microtube Composites for Oxygen Reduction Reaction Electrocatalysis in Fuel Cells

Delaney Marbach, Alvin Burns, Jack Bui, F. John Burpo, Enoch A. Nagelli*

Department of Chemistry & Life Science

United States Military Academy, West Point, New York 10996

*Corresponding PIs: Dr. Enoch Nagelli, Email: enoch.nagelli@usma.edu

High aspect ratio platinum nanotubes were synthesized using a novel bottom-up, electroless deposition approach without the aid of a template or complex deposition techniques. Square planar salt precipitates electrostatically self-assemble in aqueous solution as a function of complex ion type and concentration. Reduction of salt crystals proceeds effectively in solution using sodium borohydride reducing agent. We effectively demonstrate a cost effective and scalable approach to synthesizing platinum nanotube electrode materials with a tunable morphology and surface-area to volume ratio. We developed a platinum nanotube composite with various carbon allotrope (graphite, carbon nanotubes, and graphene) via an ink-process used for electrocatalysis at the cathode of a fuel cell. This would rival the already established electrode designs for fuel cells catalysts using platinum black, and would be a more cost-effective design. The nanotubes and inks were analyzed using scanning electron microscopy (SEM), cyclic voltammetry (CV), and linear sweep voltammetry (LSV).

527243 Size-Controlled Silver Nanoparticle Synthesis in a Jet-Mixing Reactor

Abstract:

Silver nanoparticles (Ag NPs) are of interest as catalysts for organic reactions, antimicrobial, bio-sensing and bio-imaging agents, and in conductive materials. They are conventionally produced via batch processes. These result in a wide variability in product properties such as the particle size distribution (PSD), because of poor mixing. Further, the transition of nanoparticle synthesis from lab-scale testing to bulk production calls for higher throughput, which is challenging to achieve in a batch reactor. Microfluidic reactors have been explored as alternatives that counter both these disadvantages. We illustrate the use of a simple, inexpensive continuous jet-mixing reactor for room temperature Ag NP synthesis. Ag NPs were synthesized by reduction of silver nitrate by sodium borohydride, using trisodium citrate as the capping agent. Characterization was done using UV-Visible Spectroscopy (UV-Vis), Dynamic Light Scattering (DLS) and Transmission Electron Microscopy (TEM). The effect of jet flowrate on Ag NP PSD was investigated. Increasing the flowrate from 48 mL/h to 200 mL/h indicated no change in the PSD and particle size, suggesting the reaction can be scaled-up without affecting product properties. Further, the stable long-term operation of the reactor was tested. Ag NPs sampled intermittently were found to have similar properties, demonstrating successful continuous operation. We also demonstrate that a range of particle sizes can be produced by varying the residence time between addition of reducing agent and capping agent. Future work will involve characterizing the mixing dynamics of the reactor and synthesis of bimetallic and core-shell NPs at high temperature.

528986 Mapping Evanescent Wave Scattering from Anisotropic Particles

Abstract:

Multiple experimental techniques have been developed over the past few decades to probe surface interactions in colloidal systems, which are typically a few ~kT in strength. Total Internal Reflection Microscopy (TIRM) is one such technique that measures the surface force experienced by a single spherical colloidal particle immersed in a liquid nearby, but not adhered to, a substrate. TIRM is suitable for measuring ~kT scale interactions because the technique utilizes a â€˜thermalâ€™ energy scale. Given that a broad range of colloidal particles in natural and synthetic systems are anisotropic, we are currently working to develop the conceptual framework and experimental tools to extend TIRM to anisotropic particles. Our hypothesis is that the morphology of scattered light from a particle will be a sensitive function of height and orientation. For this purpose, the scattering morphology of a spheroid at different heights and orientations is collected for sphere and ellipsoids in a map format as a standard reference for a "Scattering Morphology Resolved" - TIRM (SMR - TIRM) experiment. Results show a profound difference in the morphology of evanescent scattered light between spherical and ellipsoidal particles at various conditions. Mapping of this morphology with SMR-TIRM will be used to predict the height and orientation of the anisotropic particle at each frame, which provides an avenue to calculate the interaction energy for an anisotropic particle near a boundary.

529782 The Synthesis of Monodisperse, Supported Nanoparticle Catalysts with Switchable Surfactants and the Effects of Calcination on Nanoparticle Characteristics

Abstract:

Established methods of preparing supported nanoparticle catalysts do not provide sufficient control over nanoparticle morphology. Classical methods typically require organic surfactants to passivate the nanoparticle surface to limit undesired size changes. Stabilizing ligands bound to the nanoparticle surface compete with reagents for active sites and typically significantly hinder catalytic activity. Commonly used methods of ligand removal, such as calcination, have detrimental effects on the catalyst as exposure to high temperatures often results in significantly increased nanoparticle size and, thus, decreased total surface area for catalysis. Previously, we demonstrated a novel method for synthesizing highly active, monodisperse, supported nanoparticles using a switchable surfactant (SwiS) system. In this method nanoparticle size is finely controlled throughout synthesis and deposition. Here, we show with X-ray photoelectron spectroscopy (XPS) that supported nanoparticles prepared with SwiS are completely surface-clean after deposition, eliminating the need for any traditional activation steps such as calcination. Additionally, it is demonstrated that a low-temperature calcination at 230°C of surface-clean supported nanoparticles has detrimental effects on nanoparticle size, dispersion, and catalytic activity. Supported nanoparticles prepared with SwiS are up to 700% more active in the hydrogenation of 4-nitrophenol than their calcined counterparts. Further, calcination results in the formation of an induction time in 4-nitrophenol reduction, demonstrating that calcination causes surface structure rearrangements unfavorable for catalysis.

530140 Proximal Interactions in Graphene-Magnetic Nanoparticle Interfacial Composites

Abstract:

The single layer honeycomb carbon lattice of graphene is responsible for many of its exceptional properties such as high electron mobility, flexibility, and thermal conductivity. This has resulted in intense exploration of graphene for electronic, energy, and biosensing applications. However, the structure of graphene does not yield intrinsic magnetic order that would be desirable for potential magneto-electronic and spintronic applications. Most investigations directed toward generation of magnetism in graphene either demonstrate short range magnetism or are accompanied by disruption of its inherent properties. A recent study successfully induced long range magnetism in graphene while preserving its pristine attributes by harnessing the sensitivity of graphene electronic orbitals to proximal magnetic materials [1]. Drawing inspiration from this approach, this research examines the potential of superparamagnetic iron oxide nanoparticles (SPIONs) for proximity induced magnetism. The use of SPIONs had the potential additional advantage to locally control magnetization by directed self assembly of SPIONs. Further, the use of nanoscale materials may enable quantum level interactions. These graphene-SPION composites could thus enable emergent proximal interactions, perhaps with the ability to locally control magnetism.

Here, we describe composites that are assembled by transferring graphene to a silicon dioxide substrate onto which SPIONs have been deposited. Deposition techniques including spin coating [2] and the Langmuir Blodgett trough method [3] were utilized to produce scalable and controlled self assembly of SPIONs with different levels of ordering and surface densities. Atomic (AFM) and Magnetic (MFM) Force Microscopy were used to evaluate the topography of the composite throughout the assembly process. Subsequently, potential for perturbations in the inherent structural and electronic properties of graphene after transfer onto SPION-coated substrates was evaluated using Raman spectroscopy and resistivity measurements, respectively. To determine if magnetism was induced in the graphene structure, Hall and Quantum Hall effect measurements were conducted. Primary measurements on random, low density SPION arrays showed that composites maintained native graphene characteristics. High density arrays with ordering are being evaluated for comparison. Ultimately, this research will elucidate the potential to induce localized and global coupling of graphene to the nanomaterials in immediate vicinity.

[1] Z. Wang, C. Tang, R. Sachs, Y. Barlas, J. Shi, Proximity-induced ferromagnetism in graphene revealed by the anomalous Hall effect, Phys Rev Lett, 114 (2015) 016603.

[2] M. Benitez, D Mishra, P Szary, G. Badini Confalonieri, M. Feyen, A. Lu, L. Agudo, G. Eggeler, O. Petracic, H. Zabel, Structural and magnetic characterization of self-assembled iron oxide nanoparticle arrays, J. Phys.: Condens. Matter 23 (2011) 126003.

[3] Q. Guo, X. Teng, S. Rahman, H. Yang, Patterned Langmuir-Blodgett Films of Monodisperse Nanoparticles of Iron Oxide Using Soft Lithography, J. AM. CHEM. SOC., 125 (2003), 630-631.

530459 Probing the Kinetics of DNA-Surfactant Exchange Reactions for Carbon Nanotubes

Abstract:

Single-wall carbon nanotubes (SWCNTs) possess superior and tunable optical and electronic properties that are dependent on their helicity and handedness. Unique properties of each and every single species of SWCNTs offer possibilities in many technological applications of SWCNTs. Recent studies have shown that single-stranded DNA molecules have the ability to select structurally defined DNA-wrapped SWCNTs (DNA-SWCNTs) from their synthetic mixtures. In this work, the kinetics of DNA displacement by a surfactant for DNA-SWCNTs were measured by analyzing the spectral changes in absorption and fluorescence emission peaks of SWCNTs in the near-infrared region. The kinetics of surfactant exchange reactions were employed to characterize the binding affinity of a DNA recognition sequence for specific SWCNT species. Furthermore, the environmental impact, such as temperature, on the dispersion stability and optical properties of DNA-SWCNTs were explored. These findings will provide better understandings of sequence-dependent properties of DNA-SWCNTs and enable SWCNT applications in biosensing and imaging.

530587 Characterizing the Aqueous Dispersion of DNA-Assisted Boron Nitride Nanotubes

Abstract:

Boron nitride nanotubes (BNNTs) gained a significant attention in the scientific community due to their unique thermal, mechanical and chemical properties. However, BNNTs tend to form bundles in an aqueous medium due to their strong van der Waals interactions. As a result, the poor quality of BNNT dispersions would limit the potential liquid phase processing and applications development of BNNTs. In this work, the dispersion quality and yield of single-stranded DNA-assisted BNNTs (DNA-BNNTs) were examined against various parameters, including the ratio of DNA: BNNTs by mass, the DNA sequence using single nucleotides A_8, C_8, G_8, T_8, the DNA length using $(GT)_n$ sequences, where n=5,10,15,20, and the salt concentration in the range of 0-100 mM. The dispersion microstructure and phase behaviour of DNA-BNNTs in dilute and semidilute regime were further explored by microscopy and rheology. This will be extended to DNA-BNNT dispersions at even higher concentrations in our future studies, which will enable us to better design macroscopic assemblies of films with improved properties.

531089 Optimizing Design Parameters of a VLA-4-Targeted Liposomal Nanoparticle in a Multiple Myeloma Disease Model

Abstract:

Ligand-targeted liposomes have gathered attention as potential chemotherapeutic delivery vehicles, but inconsistencies in liposomal design have led to varying levels of success. We have systematically evaluated and discovered the significant roles that peptide-linker length, peptide hydrophilicity, peptide density, and liposome size play both *in vitro* and *in vivo*. By using a multifaceted synthetic strategy to prepare highly controlled and consistent peptide-targeted liposomes, these parameters were evaluated in a VLA-4-expressing multiple myeloma model. Studies demonstrated that increasing peptide hydrophilicity through the addition of a short oligolysine sequence while simultaneously optimizing the EG peptide-linker length led to greatly increased liposome uptake *in vitro*. A peptide-linker of length EG6 specifically lead to the greatest increase in uptake over the industry standard of an EG45 linker. *In vivo*, although peptide-targeted liposomal formulations did not lead to increased tumor accumulation compared to non-targeted liposomes, targeted liposomes demonstrated significantly increased tumor cell uptake relative to non-targeted liposome. These results show the importance of a comprehensive understanding of all facets of liposomal design on multiple biological endpoints in order to determine and properly optimize the design of a peptide-targeted liposome system.

531140 Evaluation of Mucus-Penetrating Nanocomposite Microparticles for Cystic Fibrosis-Related Infections

Abstract:

Cystic Fibrosis (CF) is one of the most common life-shortening genetic diseases, affecting more than 70,000 individuals worldwide. The median life expectancy of a patient born with CF in the U.S. is only 37 years. CF is characterized by the presence of abnormally thick and sticky mucus in the lungs and the impairment of the mucociliary clearance, which promotes persistent chronic pulmonary infections, in particular with *Pseudomona aeruginosa*as a common pathogen. The formation of biofilms is thought to protect the bacteria against antibiotic treatment, promoting drug resistance and necessitating the development of multiple antibacterial treatment options. Inhaled pulmonary delivery offers the advantage of a higher local concentration of the drug and reduced systemic side effects, as compared to those seen in oral and IV formulations. The design of PEG-coated nanoparticles offers mucus-penetrating properties that enhance their efficacy against the recurrent bacterial infections seen in CF.

The objective of this study was to develop and evaluate the efficacy of mucus-penetrating nanocomposite microparticles (nCmP) for CF-related infections, delivered via a dry powder inhaler. Polyvinyl alcohol (PVA) and Vitamin E conjugated to 5kDa poly (ethylene glycol) (VP5K) coated polymeric nanoparticles (NP) were synthesized via single emulsion using acetalated dextran (Ac-Dex), a biodegradable and tunable polymer. Both formulations resulted in spherical, smooth, highly homogeneous nanoparticles with a diameter of ~200 nm and a slightly negative Î¶ potential. The nCmP were produced via spray drying and preliminary characterization showed optimal size and distribution throughout the lungs, evaluated by a Next Generation Impactor (NGI).

Experiments are ongoing, with the primary objective of comparing VP5K and PVA NP to determine their mucus-penetrating properties in artificial mucus and their particle transport rates via particle tracking analysis. Curcumin is used as a drug model because of its fluorescence and known antibacterial properties. Secondary outcomes for this study will include antibacterial efficacy against *P. aeruginosa* PA01, a common microbe seen in CF-related pulmonary infections. Overall, VP5K-loaded-NP are a promising alternative for rapid diffusion through the mucus barrier, to release the antibiotic for treatment of *P. aeruginosa*infection in CF patients.

532155 Nanoharvesting and Nanodelivery of Bioactive Materials Using Engineered Silica Nanoparticles

Abstract:

Functionalized mesoporous silica nanoparticles (MSNP) possess surfaces that can be readily modified with specific functional groups for targeted binding interaction, in addition to ample pore space for loading of bioactive materials to be transported through cellular walls and membranes. Engineered silica nanoparticles (ESNPs) have been used extensively to deliver bio-active materials to target intracellular sites (nanodelivery). These materials have been fairly well explored as non-viral vectors for nucleic acid (DNA/RNA) delivery such as for siRNA induced interference. On the other hand, the reverse process guided by the same principles is called â€œnanoharvestingâ€, where valuable biomolecules are carried out and separated from living and functioning organisms using nanocarriers, which are designed to enter the cells and be expelled without inducing significant toxicity. This work focuses on the interaction (uptake/secretion) of ESNPs with eukaryotic cells especially lipid membrane penetration and intracellular localization to achieve a common ESNP design principles for both nanoharvesting and nanodelivery.

Nanoharvesting of polyphenolic flavonoids (a model class of therapeutics) from plant hairy roots is demonstrated using ESNPs (mean particle diameter ~170 nm) in which titania and amine functional groups are used as specific binding sites and a source of positive surface charge source, respectively. Intracellular uptake and localization of fluorescently tagged particles inside of hairy root cells are visualized by fluorescent microscopy, and isolation of therapeutics is confirmed by increased pharmacological activity of solutes recovered from the particles. After nanoharvesting, roots are found to be viable and capable of therapeutic re-synthesis, suggesting non-toxic nature of the designed ESNPs. The combination of both amines and titania binding sites is found to be necessary for uptake and metabolite binding, respectively.

In order to identify the underlying nanoparticle uptake/expulsion mechanism, the titania content of the plant roots was quantified as a function of time with exposure to nanoparticles. Fluorescence uptake and particle exchange experiments, in which fluorescently labeled ESNPs were taken up by plant cell cultures and then recovered in solution were also used to quantify ESNP uptake and recovery. Temperature (4 °C and 23 °C) dependent experiments showed similar uptake and recovery for both temperature indicating an energy independent process (inactive transport) for this particle size and surface functionalization.

While seemingly diverse, the highly tunable nature of ESNPs and their interactions with eukaryotic cells are broadly applicable, and enable facile nanodelivery based on a continuous uptake-expulsion mechanism.

533433 Nanoclustering of Salicylic Acid in Organic Solvents

Abstract:

In crystallization, the primary step is nucleation, which is a first-order phase transition where solid particles are formed from their components. Nucleation in past was explained with the help of classical nucleation theory (CNT). Nano-clustering explained through two-step nucleation theory (TNT), challenges classical nucleation theory (CNT) as it had shortcomings. The difference in the two theories are in terms of structural and density fluctuations. In CNT, they occur simultaneously however in TNT density fluctuations follow structural fluctuations.[1] Thus, the nanoclusters are highly dynamic in nature, which rearrange themselves with time. In recent times, researchers have found evidences that the theory exist in various proteins[2] and inorganic particles[3], by means of experiments and molecular simulations. In addition, the role of solvent is an important aspect for solute-solute interactions.

My research focuses on the study of nanoclusters formed in salicylic acid as a function of time in different organic solvents. Salicylic acid, an anti-fungal and anti-inflammatory drug, is used in cosmetic industry. With the use of dynamic light scattering, nanoclusters of salicylic acid were studied in ethyl acetate, acetonitrile, methanol and acetic acid solutions at different concentrations at an isothermal condition for a period of 72h. The solutions are undersaturated, saturated and supersaturated at 25°C. With the change in concentration in different solvents, a trend in order of clustering for different solvents was observed, which holds good for the prediction of nucleation order obtained earlier[4]. At all times, the increase in size is observed which states the order of clustering in different solvents. However, these trends tend to change over the period of time. The one with lower cluster size in undersaturated solutions obtains higher cluster size when the solutions are supersaturared however vice versa is also found to be true. In all solvents, a similar size is obtained at saturated concentration.

References:

[1] D. Gebauer, M. Kellermeier, J. D. Gale, L. Bergström, and H. Cölfen, "Pre-nucleation clusters as solute precursors in crystallisation.," *Chem. Soc. Rev.*, vol. 43, no. 7, pp. 2348–71, 2014.

[2] P. G. Vekilov, "Nucleation of crystals in solution," *AIP Conf. Proc.*, vol. 1270, pp. 60–77, 2010.

[3] D. Gebauer, A. Völkel, and H. Cölfen, "Stable prenucleation calcium carbonate clusters.," *Science*, vol. 322, no. 5909, pp. 1819–22, 2008.

[4] D. Mealey, D. M. Croker, and A. C. Rasmuson, "Crystal nucleation of salicylic acid in organic solvents," *CrystEngComm*, vol. 17, no. 21, pp. 3961–3973, 2015.

533964 Rapid Photo-Actuation of a DNA Nanostructure Using an Internal Photocaged Trigger Strand

Abstract:

DNA nanotechnology has emerged as one of the most powerful methods for constructing nano-mechanical devices with precise addressability. The ability to actuate these devices on-demand would provide a powerful method for creating dynamic nanomachines that can reconfigure nanostructures, apply precisely defined forces, or spatiotemporally control self-assembly. Typically, DNA nanostructures have been temporally modulated by either 1) introducing external toehold-mediated displacement strands, or 2) incorporating photoswitchable azobenzene nucleotides that can reversibly break hybridization. However, these approaches suffer from drawbacks like 1) the need to introduce an exogenous strand (which is not feasible for many applications, especially in biology), and 2) slow kinetics or undesired reversibility. To circumvent these limitations, we have developed a series of photo-caged oligonucleotides that block hybridization until illuminated with UV light. By incorporating caged displacement strands within an existing nanostructure, we can prevent their action until illuminated. These â€œspring-loadedâ€ nanostructures are in a metastable state until activated by light, whereupon the high local concentration of the displacement strand will drive irreversible nanostructure reconfiguration. To demonstrate this concept, we designed a nano-mechanical tweezer that switches between the closed and open state with UV light. In effect, this device can apply a nano-mechanical force within a few seconds of illumination, paving the way for dynamic nanomachines that can exert controlled motion. The tweezer was analyzed using AFM, FRET, and computational simulations. Our results demonstrate that the photocaged mechanism is efficient and fast, surpassing externally added strand displacement by almost two orders of magnitude. We envision that this approach will allow for light-activated nanomachines in the future for biophysical studies with cells.

The tweezer was analyzed using a number of mechanisms, including single-molecule spectroscopy, and will highlight their fast switching kinetics, biocompatibility, and efficiency. We envision that photocaged internal displacement strands will find broad applicability for switchable nanostructures (e.g. for drug delivery), spatiotemporally-controlled self-assembly, or biophysical studies such as exerting controlled forces on proteins.

533966 Targeted Delivery of a Drug Coupled Gold Nanoconjugate Induces Respiratory Recovery Following Cervical Spinal Cord Injury in Rats

Abstract:

The gold nanoconjugates consists of a spherical gold nanoparticle drug carrier chemically conjugated to methylxanthine drugs, and a targeting protein (wheat germ agglutinin coupled to horseradish peroxidase or WGA-HRP). In this poster, we demonstrated a strategy of selectively targeting the lower motor (phrenic) and pre-motor rostral ventral respiratory group (rVRG) neurons responsible for diaphragmatic function in rats by showing persistent respiratory recovery 48 hours after a single intradiaphragmatic injection of the nanoconjugate at 0.1% of free drug dosage inducing therapeutic effect. Biodistribution study was conducted by determining the amount of gold in the blood, RES organs, and targeted tissues after injection. Inductively coupled plasma mass spectrometry was used in this study. The targeted transportation was also demonstrated by anti-WGA immunohistochemical analysis. Respiratory recovery data will include electromyography (EMG) analysis, phrenic nerve recordings, and whole body plethysmography (WBP) on a validated animal model. We also carried out in-vitro drug release studies by monitoring drug released from the nanoconjugates incubated in buffers mimicking different circumstances for time periods ranging from 1 hour to 4 weeks by LC-MS/MS. To conclude, a single administration of the nanoconjugate restored 75% of the respiratory drive at 0.1% of the effective systemic drug dose. This work has important translational implications.

534607 The Implications of Competitive Adsorption on Lipoprotein-Nanoparticle Biodistribution

Abstract:

Nanoparticles introduced into the bloodstream are exposed to thousands of proteins that are competing for their surfaces. During transport, this competition changes the amount and class of proteins that are bound to the nanoparticle surface. High density lipoproteins (HDLs) have been shown to bind nanoparticle surfaces of different sizes and chemistries, but quantitative values are lacking. Bound HDLs could change the biodistribution of nanoparticles if the interaction is stable and HDLs retain their functionality. We determined equilibrium binding parameters for HDLs and nanoparticles using gel electrophoresis and a quantitative model of protein-sphere interactions. We used polystyrene spheres - pristine, carboxylate-modified, and amine-modified - as model nanoparticles. ApoAI â€“ the main structural protein of HDLs - had the following affinities for the nanoparticles: $K_{d\text{-}COOH}$ ~1.1 ÂµM, $K_{d\text{-}NH2}$ ~1.9 ÂµM, $K_{d\text{-}PS}$ ~3.1 ÂµM). Interestingly, although there is relatively strong binding of ApoAI to $PSNH_2$, ApoAI does not completely saturate the nanoparticle surface at constant plasma concentration and increasing nanoparticle concentrations. These results provide insight into the complexities involved in the binding kinetics of lipoproteins to nanoparticles and the need to flesh out the possible involvement of lipoproteins in redirecting nanoparticle drug delivery vehicles to HDL receptors on the liver.

534846 Development of Interfacial Mechanical Strength for Armored Gas Filled Capsules

Abstract:

The fluid-fluid interface of an air bubble or oil droplet can be stabilized through the adsorption of surface active particles. If unconfined, the bubbles and droplets will assume spherical shapes to reduce their surface area and thus the interfacial energy. Within a confined geometry, adsorption and jamming of particles on an interface can lead to the retention of non-spherical shapes. Here, we examine the role of particle adsorption and interfacial compression in forming armored non-spherical air filled capsules. Capsules are generated by flowing an air bubble through a capillary filled with a suspension of surface active particle â€“ surfactant complexes. Our work uses colloidal suspensions of silica nanoparticles coated with cetyltrimethylammonium bromide (CTAB). The complexes adsorb to the interface and are convected to the trailing end of the bubble, armoring the surface. The interface is compressed as the bubble exits the capillary resulting in the retention of a non-spherical shape. Time scale estimates and controlled interfacial measurements for a pinned air bubble are used to characterize particle adsorption to the interface. Controlled compressions are applied, and the interfacial rheology is compared before and after compression. Results from these experiments are combined with observations of changes in bubble shape in situ to characterize adsorption and compression during capsule generation. These results provide insight into the evolution of the interfaceâ€™s mechanical strength to form the armored air-filled capsules.

535263 A Colorimetric Sensor for the Detection and Quantification of Therapeutic Levels of Ionizing Radiation

Abstract:

Radiotherapy is a medical procedure that uses high energy photons for tumor ablation. Evolution of radiotherapy has led to minimization of side effects in patients. To ensure an enhanced quality of life for the patient, there is a need to monitor and record the prescribed dose delivered to the patient. Conventional sensors are complex and require skilled operators to handle them. This complexity involved in existing sensors make their use limited in the clinic. To overcome these concerns, we have developed a colorimetric sensor that can detect therapeutic levels of ionizing radiation. Metal salt in the presence of radiation undergo reduction to form metal atoms. These metal atoms nucleate and grow to form stable nanoparticles which leads to the development of characteristic color depending on the shape and size of the particles. The resulting is color is quantified using Ultra-Violet Spectroscopy and used as an indicator for the delivered radiation dose. A hydrogel based sensor was designed to alleviate handling concerns. We successfully demonstrated the clinical applicability of the sensor by measuring the dose delivered to a canine patient undergoing radiotherapy. Our predictions match the prescribed dose making our sensor a potential candidate in the future to be used in routinely in the clinic.

535649 Probing Nanoclustering of Fenoxycarb in Isopropanol Solutions

Abstract:

Crystal nucleation is a widely occurring phenomenon in nature, *e.g.* in the formation of bones, shells and ice, is of significant industrial importance in the manufacturing of many inorganic and organic materials, and has medical implications, *e.g.* in the formation of kidney stones and precipitates of amyloid proteins. Despite its importance, the current molecular level understanding of the mechanisms of crystal nucleation from solution is poor, and consequently, there are features of crystal nucleation that still lack proper explanation, *e.g.* the so-called â€œhistory of solutionâ€ effect and the influence of filtration on nucleation. Classical theory of nucleation originally developed early in the last century[1] has been used for years to understand primary nucleation. Conventionally crystallization is assumed to proceed via the attachment of ions/molecules to a primary molecule. In recent years, experimental evidence for the existence of various kinds of clusters without crystalline structure in both supersaturated and undersaturated solutions, with sizes of the order of ~100 nm, has started to accumulate[2-7]. This has led to development of alternative nucleation theories[8, 9], which take their point of origin in the various concepts of clusters. These clusters are proposed to be thermodynamically stable entities without a defined phase boundary, which can be present also in undersaturated solutions. According to the *two-step theory* of nucleation, originally proposed for nucleation of proteins, concentration fluctuations result in the nucleation of a kind of clusters having a liquid-like structure with a higher concentration of the solute compared to the bulk solution. Final crystal nucleation then takes place within the clusters[10, 11]. The extent of applicability of these models, and the underlying molecular level mechanisms, are still focal points of investigation and debate.

In the present work, we have investigated the molecular association and clustering of fenoxycarb molecules in isopropanol solution at room temperature with DLS (Dynamic Light Scattering, Malvern Zetasizer ZSP Nano), and verified further using SAXS (Small Angle X-ray Scattering, I22 beamline, Diamond Light Source, UK) and NTA (Nanoparticle Tracking Analysis, Malvern Nanosight LM14). As opposed to most of the molecules so far featured in studies on clustering in solution the fenoxycarb molecule is bigger and have conformational flexibility. Fenoxycarb solutions of various concentrations (x) were very carefully prepared, systematically and periodically analysed with DLS, SAXS and NTA at 25 °C. The presence of clusters in isopropanol solutions of fenoxycarb has been clearly established in this work, using three different analytical techniques. Clusters are found in both under- and supersaturated solutions. There is a good correspondence between the three analytical techniques in terms of cluster size, which ranges up to several hundred nanometers at undersaturated conditions and a few micrometers at supersaturated conditions. The cluster size is found to increase with solute concentration at constant residence time. The observations of the NTA and SAXS experiments on solutions pre-treated at an elevated temperature are consistent with experimental observations of primary nucleation under different conditions, and help explain the strong solution history effects exhibited by this system[12].

Acknowledgments: *We gratefully acknowledge the financial support of the Synthesis and Solid State Pharmaceutical Centre, funded by Science Foundation Ireland (grant no. 12/RC/2275), as well as support from the Bernal Institute and the Department of Chemical Sciences, University of Limerick.*

References:

[1] Becker, R. and Döring, W., *Ann. Physik, **416**, 719 (1935).*

[2] Gebauer, D., Völkel, A. and Cölfen, H., *Science **322**, 1819 (2008).*

[3] Tanaka, S., Ito, K., Hayakawa, R. and Ataka, M. J., *Chem. Phys. **111**, 10330 (1999).*

[4] Sorensen, T. J., *Chem. Eng. Technol. **37**, 1959 (2014).*

[5] Sorensen, T. J., Sontum, P. C., Samseth, J., Thorsen, G. and Malthe-Sorenssen, D., *Chem. Eng. Technol. **26**, 307 (2003).*

[6] Chattopadhyay, S., Erdemir, D., Evans, J. M. B., Ilavsky, J., Amenitsch, H., Segre, C. U. and Myerson, A. S., *Cryst. Growth Des. **5**, 523 (2005).*

[7] Jawor-Baczynska, A., Sefcik, J. and Moore, B. D., *Cryst. Growth Des. **13**, 470 (2013).*

[8] Vekilov, P. G., *J. Cryst. Growth **275**, 65 (2005).*

[9] Gebauer, D., Kellermeier, M., Gale, J. D., Bergström, L. and Cölfen, H., *Chem. Soc. Rev. **43**, 2348 (2014).*

[10] Pan, W.; Kolomeisky, A. B.; Vekilov, P. G. *J Chem Phys* **122**, 174905 (2005).

[11] Vekilov, P. G. *Cryst. Growth Des.* **4**, 671 (2004).

[12] Kuhs, M.; Zeglinski, J.; Rasmuson, Å. C. *Cryst. Growth Des.* **14**, 905 (2014).

535919 Effects of Silica Nanoparticles in PVDF-SiO_2 Mixed Matrix Membranes Developed ViaÂ Immersion Precipitation Phase Inversion

Abstract:

Membranes have large surface areas ideal for loading inorganic nanoparticles (NPs). Polyvinylidene Fluoride (PVDF) and Silica (SiO_2) can be used as a dope solution to form a SiO_2 -PVDF membrane via phase inversion. The ease and diversity in surface properties allow for the design of SiO2 mixed matrix membranes (MMMs) with specific functions and properties for a wide range of applications and scalable processes. In the current study SiO_2 -PVDF MMM were developed via immersion precipitation phase inversion using a SiO_2 wt% of 2,4,6 and 8. The MMMs exhibited an increase in flux with increasing SiO_2 wt% with the highest having a measured flux of 0.58 ± 0.01 (LMH) at 5 bar. Also, the porosity and mean pore radius increased as the silica volume fraction increased. Indicating that if not well dispersed, the addition of SiO_2 nano-particles will cause defects around the aggregates, thus resulting in a change in the mean pore radius and flux.

535975 Carbon Black Morphology, Light Scattering and Direct Radiative Forcing

Abstract:

Carbon black (soot) nanoparticles generated by combustion sources involving transportation, power generation and fires form fractal-like, porous clusters (agglomerates).[1] Yet, their optical properties are calculated using the Mie theory for spheres neglecting the ramified agglomerate morphology and impeding the accurate estimation of carbon black environmental impact, characterization by optical diagnostics and selective detection with fire sensors.

Here, the carbon black morphology and radiative properties are investigated experimentally and simulated during surface growth and agglomeration by Discrete Element Modeling (DEM)[2] coupled with Discrete Dipole Approximation (DDA).[3] The DEM-derived number of constituent primary particles (PPs), effective density and differential scattering cross-sections for vertically-, C_v, and horizontally-polarized incident light, C_h, are in excellent agreement with those measured here in different premixed ethylene flame conditions. In contrast, the Rayleigh Debye Gans (RDG) theory coupled with a scaling law for agglomerates with monodisperse, single, non-aggregated PPs[4] underestimates the DEM-derived C_v and C_h by 60 %. The integral mass scattering (*MSC*) and absorption cross-sections (*MAC*) estimated by DEM-DDA account for the detailed carbon black fractal-like morphology, yielding average direct radiative forcing of 0.70 ± 0.07 W/m^2. This is 45 % smaller than the maximum global climate model predictions using the Mie theory for spheres. The C_v, C_h, *MSC* and *MAC* derived by RDG theory coupled with a DEM-derived scaling law accounting for PP aggregation and polydispersity[1] are in good agreement with those estimated by DDA. Thus, both DEM-DDA and the revised RDG theory can be used for the optimization of climate forcing estimations and carbon black optical diagnostics, as well as for selective detection by fire sensors.

References:

[1] Kelesidis, G. A.; Goudeli, E.; Pratsinis, S. E., Morphology and mobility diameter of carbonaceous aerosols during agglomeration and surface growth. *Carbon* **2017**, *121*, 527-535.

[2] Kelesidis, G. A.; Goudeli, E.; Pratsinis, S. E., Flame synthesis of functional nanostructured materials and devices: Surface growth and aggregation. *Proc Combust Inst* **2017**, *36*, 29-50.

[3] Kelesidis, G.A.; Pratsinis, S.E., From nascent to mature soot light absorption during agglomeration and surface growth. *Proc Combust Inst* **2018**, *under review*.

[4] Sorensen, C. M., The Mobility of Fractal Aggregates: A Review. *Aerosol Sci Technol* **2011**, *45*, 765-779.

536274 Chitosan Electrospun Nanofibers Functionalized with Collagen By Carboxamide Bond Formation

Abstract:

Natural polymers are biomaterials of high biotechnological value. Their biodegradability, biocompatibility and high availability are some attributes that make them attractive materials for multiple innovative applications. They have been object of study for a wide range of novel biomedical applications due to their biocompatibility with living tissues. Among other polymers, chitosan and collagen are very attractive in the biomedical field.

Chitosan is an amino-polysaccharide derived from chitin, which is particularly important as a promoter of tissue regeneration; it has hemostatic and immunomodulatory properties and a wide antimicrobial and antifungal potential. Chitosan has been combined with polyethylene glycol, polyvinylpyrrolidone, gelatin, polyvinyl alcohol and especially collagen to produce hybrid materials. Whereas, collagen is one of the most important and abundant biopolymers with high biocompatibility and particular characteristics that distinguishes from synthetic polymers. Its abundance and structural properties confers to collagen a wide range of biomedical applications, especially in tissue engineering. Collagen directly affects the morphology and physiology of connective and epithelial tissue cells, and has great potential to be use as basis for scaffolds and biomaterials production; it is well know that it can interact with living tissue and provides adequate biomechanical and biochemical environments for living tissue.

Because of the above mentioned, chitosan and collagen have inspired the development of hybrid materials, and different procedures had been widely studied in order to obtain potentially useful materials focused on tissue engineering. Chitosan/collagen combination does not occur by itself in nature; nonetheless, this combination is very attractive and suitable to generate hybrid advanced biomaterials with unique mechanical and biomedical properties. The versatility of these polymers to be processed into different structural shapes has created high expectation for the development of novel functional materials, and electrospun nanofibers are of especial interest. There are several reports about the exploration of electrospining conditions and parameters to produced chitosan and collagen based nanofibers, however, the processability of chitosan through this technique, to obtain collections of nanofibers of good quality requires the use of additives or harmful solvents, such as trifuoroacetic acid (ATF), and the produced nanofibers are highly soluble. While in the case of collagen, besides its low spinnability, it has been reported that the electrospinning process can significantly affect the bioactivity of native collagen reducing its regenerative potential. To address these problems, this work presents the surface functionalization of chitosan electrospun nanofiber mats with collagen, as a better alternative to include collagen into a composite electrospun biomaterial preserving the bioactivity. Taking advantage of the amino groups of chitosan and the carboxylic acid groups of collagen as binding sites under precise conditions, post-electrospun coupling is feasible via carboxamide bond formation between both polymers promoted by EDC/HOBt. This strategy allows the achievable inclusion of collagen on chitosan nanostructures without affecting the bioactivity of both polymers.

536900 Electrospun Nanofibers from a Blend of Asphaltenes with Cellulose Acetate

Abstract:

In this work, we report the successful fabrication of nanofibrous mats prepared from a blend of asphaltenes with cellulose acetate. Asphaltenes are the components of crude oil with the highest molecular weight. They are an inexpensive feedstock, consisting of polycyclic aromatic hydrocarbon rings with attached aliphatic chains. Their presence in the crude oil is associated with processing difficulties caused by their precipitation during transportation and refining. However, they are suitable as carbon fiber precursors due to their high carbon content (>80%). On the other hand, cellulose acetate is derived from the processing of cellulose, the most abundant natural polymer on earth. It is therefore a low-cost material derived from a biorenewable resource.

The concept behind this study is to combine these inexpensive resources for the manufacture of high added-value products. In this context, we prepared nanostructured fibrous mats from a blend of asphaltenes with cellulose acetate via the electrospinning technique. With the aid of a design-of-experiments statistical methodology, the process was optimized and the average fiber diameter was minimized. The minimization of the average fiber diameter, had a large impact on applications where the high surface area is important, such as in adsorption, in catalysis and in filtration. Moreover, we report the characterization results of the electrospun nanofibers based on their thermal, morphological and structural properties. Hence, optimal conditions can be obtained for predetermined diameters and desired morphologies of the nanofibers.

536904 Massive Enhancement of Optical Transmission across a Thin Metal Film Via Wave Vector Matching in Grating-Coupled Surface Plasmon Resonance

Abstract:

Investigations in grating coupled surface plasmon resonance often use only single mode resonances. In this study, we demonstrate the coupling of distinct surface plasmon resonance modes on opposite sides of a thin film metallic grating by varying the refractive index (RI) on either side of the grating. This coupling occurs by matching the resonance wavelengths, a function of the refractive index, across metal layer. In the first case, we deposited a high refractive index (RI) SnO_2 of increasing thickness to red shift the front side surface plasmon until it couples with the backside surface plasmon across a semi-transparent ~45 nm thin silver grating. Over an order of magnitude increase in transmission enhancement is shown, from 5-fold transmission enhancement through a bare silver grating referenced with flat silver to 100-fold enhancement after deposition of a tin oxide layer of appropriate thickness (~310 nm). This optical transmission enhancement is then probed through analytical experiments with liquids of various refractive indexes and computational studies. This matched system shows an increase in intensity sensitivity and decrease in wavelength sensitivity with respect to RI change. The deposited SnO_2 on the front side avert access of analytes to the enhanced electric field on the silver surface. Therefore, we also demonstrated an alternative way to couple the two plasmon modes. In the second case, a lower refractive index substrate was used to blue shift the back side surface plasmon and couple it to the front side surface plasmon by index matching using aque-ous solution of various RIs. For aqueous based matched system, the transmission enhancement increases from 3-fold for a bare silver grating to 13-fold for matched system.

536985 PVA â€“Based Nanostructuraed Catalysts Support Functionalized with Pyrrolylquinone-Tetrazole

Abstract:

The Organocatalysis has been of great importance in the last years and has focused on the development of efficient organic catalysts to get high yield to selectivity ratio and to promote transformations using substoichiometric quantities. Particularly, the selective oxidation of organic compounds is a great challenge in chemical synthesis, and active organic molecules are a feasible alternative as efficient catalyzers in the oxidation of organic compounds. In this context, the quinones are excellent agents for electrons transference and stand out as catalysts in oxidation reactions.

Recently, a new pirrolilquinone based molecule has demonstrate a great potential as organic catalyst in oxidative transformations. However, the catalysts recovery and recycling is often complicated, and has been a constant challenge to develop sustainable processes and strategies. The Immobilization of organocatalysts in polymeric nanostructured supports has been an outstanding strategy in the last years to overcome the problems associated to catalysts recovery and recycling. In addition, the high volume to surface area ratio of nanofiber supports provide a higher contact area, mimicking thus the homogeneous catalysts. The nanostructured polymeric supports offer several active sites; therefore, the organocatalyts can be incorporated in large quantities along the polymeric support, giving a wide variety of applications according to the reaction conditions.

Polyvinyl alcohol (PVA) is a highly attractive polymer for the production of nanostructured supports. Its physicochemical properties make PVA a versatile polymer for its processing through the electrospinning technique, obtaining high quality nanofiber collections. PVA can be functionalized through the formation of covalent bonds between the hydroxyl groups of PVA and the ligand through substitution or esterification reactions. This work present the synthesis of a polymeric nanostructured support for the immobilization of pyrrolylquinone-tretrazole organocatalyts, using the multicomponent reaction strategy Ugi-Azida; a sustainable and environmentally friendly approach.

537301 Investigations into the Generation of Chitin Nanofibers By Cryogenic Grinding

Abstract:

Chitin, second only to cellulose, is a highly abundant biopolymer found in nature, but unfortunately the majority of it is just discarded as commercial waste. A chitin fiber matrix is the primary component of the exoskeletons of most arthropods, including insects, crabs, and prawns along with some minerals and proteins. Even after the removal of the minerals and proteins, the chitin fibers are still strongly held together by hydrogen bonding. Chitin is also not soluble in most common solvents. This can make the fibrillation process of chitin quite difficult. Although chitin nanofibers are not yet commercially available, there are currently several methods to generate them mentioned in the literature including high-pressure homogenization, grinding with specially designed grinding stones, and electrospinning. New methods and further advancements to the existing methods of chitin fibrillation are still needed. Preliminary investigations into the generation of chitin nanofibers by cryogenic grinding have only shown limited success. In this study, commercially purchased chitin was ground using a cryogenic grinder and additional modifications were made to the process to further improve fibrillation. Characterization of thermal and morphological properties of cryo-ground chitin as well as its synergistic incorporation into biodegradable polymers are explored and discussed in this work.

537358 Study of ZIF-8 MOF's as Viable Drug Carriers

Abstract:

Metal-organic frameworks (MOFs) are compounds consisting of metal ions or clusters coordinated to organic ligands forming one-, two-, or three-dimensional structures. MOFs are considered a subclass of polymer clusters with reticular structures. There are more than. 20,000 different reported MOFs within the last decade, and their properties vary because of the nature of the metal compounds (ions or clusters) and ligands[4].

The most common used ligands or organic units are ditopic or polytopic organic carboxylates, imidazoles, or other molecules with active heteroatoms. When linked to metal-containing units, yield architecturally robust crystalline MOF structures with a typical porosity of greater than 50% of the MOF crystal volume. The surface area values of MOFs typically range from 1000 to 10,000 m2/g, thus exceeding those of traditional porous materials such as zeolites and carbons[4]. To date, MOFs are more extensive in their variety and multiplicity than any other class of porous materials. Generally these frameworks are used for absorption or adsorption of different gases, such as methane, CO_2 and hydrogen. Taking advantage of their large contact area and chemical tuning, MOFs are also used for catalysis and as sensors or indicators for gases and disperse components in solution[4].

Recent studies are facing to use MOFs in medical applications such as drug transportation. Target molecules can be stored, transported and delivered in a specific and controlled manner at precise locations without modifying or disrupting the purpose or function of the bioactive molecules. Reports show the use of MOFs and other nano-materials for direct drug administration, stabilization of the drug through encapsulation or surface attachment, aid in cellular internalization, targeting delivery to a specific cell population, and providing controlled release of the drug at the designated target[3]. Organic-based drug delivery systems (DDSs) including lipid- and polymer-based systems which are already approved for clinical use[1]. The use of inorganic nanomaterials such as MOFs in biological research represents one of the fastest growing areas of interest. Some examples of the effective use of the MOFs in DDSs with reported viability are ZIF-6 in anticancer drugs delivery and transport. Such as MOFs MIL-100 and MIL-101 achieving absorption of ibuprofen and the MOFs ZJU-101, MOF-74-FE(III) and Bio-MOF-1 tested to load diclofenac and deliver it to specific inflamed areas[2,3].

In this work zinc-2-methylimidazole framework (ZIF-8) is used as encapsulation system for different bioactive molecules. Stability of the clusters was analyzed in acid media[6]. ZIF-8 is a nontoxic molecule built from a salt of zinc and 2-methylimidazolate as organic ligand. This MOF has pores of 3.4 Å of diameter and a cavity of 11.6 Å of diameter. This material is stable under physiological conditions that gives it the potential of being an effective storage for bio-active molecules[3].

The bio-active molecules were provided by the "Bioactive Molecule Synthesis Research Group" of the La Salle University Mexico. The molecules comprise a family of derivatives of Pyrazino[2,3-b]quinoline-10(5H)-one (PZQ) with different functional groups. Zinc nitrate hexa-hydrated and 2-methylimidazole were purchased from chemical distributors and used without further purification. Encapsulation is performed using one-pot synthesis[4] by mixing a solution of the Bio-active molecule and 2-methylimidazole in methanol whit the solution of zinc nitrate hexahydrate in methanol at room temperature. The solution is mixed for 30 minutes and then centrifuged at 1500 RPM for 10 minutes to remove the supernatants. The material is washed with DI water in three consecutive cycles and dried in the oven at 50°C for 3 hours.

The efficiency of encapsulation has been determined by measuring the bioactive molecules in the supernatant of the reaction mixtures using UV-Vis spectrophotometry and FT-IR. Solid structures have been characterized by FT-IR and X-ray diffraction. No important difference in microstructure is observed between ZIF-8 alone and the MOF with the encapsulated bioactive molecule; which we assume there is no chemical interaction between cages and encapsulated molecules.

Stability of the cages has been analyzed in different pH systems observing the MOF disintegrates in strong base and acid media which implies that biomolecules can be delivered in the gastric system. Studies of biocompatibility and acute toxicity are being carried out to determine the viability of the use of these materials as drug delivery systems.

References:

[1] Emmanuel N. Koukaras, Tamsyn Montagnon,Pantelis Trikalitis, Dimitrios Bikiaris, Aristides D. Zdetsis, and George E. Froudakis. «Toward Efficient Drug Delivery through Suitably Prepared Metal–Organic Frameworks: A First-Principles Study.» *The journal of Physical Chemistry*, 2014: 8885-8990.

[2] Jia Zhuang, Chun-Hong Kuo, Lien-Yang Chou, De-Yu Liu, Eranthie Weerapana, and Chia-Kuang Tsung. «Optimized Metal-Organic-Framework Nanospheres for Drug Delivery: Evaluation of Small-Molecule Encapsulation.» *ACSNANO*, 2014: 2812-2819.

[3] Kyo Sung Park, Zheng Ni, Adrien P. Côté, Jae Yong Choi, Rudan Huang, Fernando J. Uribe-Romo, Hee K. Chae,. «Exceptional chemical and thermal stability of zeolitic imidazolate frameworks.» *PNAS*, 2006: 10186-10191.

[4] Paolo Falcaro, Raffaele Ricco, Cara M. Doherty, Kang Liang, Anita J. Hillb and. «MOF positioning technology and device fabrication.» *Royal Society of Chemistry*, 2014: 49.

[5] Patricia Horcajada, Christian Serre, Guillaume Maurin, Naseem A. Ramsahye,Francisco Balas, Marı´a Vallet-Regí, Muriel Sebban, Francis Taulelle, and Gérard Férey. «Flexible Porous Metal-Organic Frameworks for a Controlled Drug Delivery.» *JACS articleas*, 2008: 6774-6780.

[6] Yanyu Yang, Quan Hu, Qi Zhang, Ke Jiang, Wenxin Lin, Yu Yang, Yuanjing Cui, and Guodong Qian. «A Large Capacity Cationic Metal-Organic Framework Nanocarrier for Physiological pH Responsive Drug Delivery.» *American Chemical Society*, 2016: 17.

538996 Can Nanotechnology Land a Solution for the Energy Security Challenge?

Abstract:

Energy sustainability is essential for the continuity of the societyâ€™s modern life and their further development. The energy demands rising as a result of the worldwide population growing while fossil fuel is yet the main energy resource to meet this high demand. In 2016, many countries have signed the Paris agreement to reduce the global warming elevated levels by using industrial technology-driven innovations and get the world ready for the new energy era. Nanotechnology applications in the energy sectors may land a solution for the energy security challenge. We have synthesized several inexpensive and scalable novel nano-platforms for a number of energy applications. Examples include converting the less-valuable materials into lighter clean fuel and boosting oil recovery beyond waterflood. In this talk, we will share examples of these smart nanomaterials and the lights will be shed on their synthesis, characterization and contribution to the field.

536439 Synthesis and Characterization of Biogenic Selenium Nanoparticles with Antibacterial Properties

Abstract:

Statement of Purpose: Antimicrobial resistance is a global concern that affects to more than 2 million of people each year. Therefore, new approaches to kill bacteria are needed. One of the most promising methodologies may come from metallic nanoparticles since bacteria cannot develop resistance to these nanostructures. While metallic nanoparticle synthesis methods are well studied, they are often accompanied by significant drawbacks such as cost, extreme processing conditions, and toxic waste production. In this work, we explored the environmentally safe synthesis selenium nanoparticles, which have shown promise in killing bacteria.

Methods: Using*Escherichia coli*(strain K-12 HB101; Bio-Rad, Hercules, CA), *Pseudomonas aeruginosa*(Schroeter, Migula, ATCC® 27853™), *Staphylococcus aureus*subsp. aureus Rosenbach (ATCC® 12600™) and Methicillin-resistant *Staphylococcus aureus*(MRSA) (F-182, ATCC® 43300™) were used to synthesize biogenic selenium nanoparticles, 90-150 nm selenium nanoparticles were synthesized under standard conditions using an environmentally-safe approach. Bacterial cultures, prepared in Luria-Bertani (LB) media, were inoculated with 2 mM sodium selenite (Na2SeO3) solution to activate the detoxification process that leads to the synthesis of nanoparticles. The nanostructures were characterized using Transmission Electron Microscopy (TEM, JEOL USA Inc., MA), Energy Dispersive X-Ray (EDX) to determine the chemical compositions and Inductively coupled plasma mass spectrometry (ICP-MS) to validate the chemistry within the samples. Nanoparticles were also characterized and tested for their ability to inhibit the bacterial growth through optical density measurements in a SpectraMax M3 spectrophotometer (Molecular Devices, Sunnyvale, CA), and colony forming unit assays. Besides, biocompatibility tests of the nanoparticles with human tissue were accomplished, growing human dermal fibroblast (HDF) cells in media in the presence of biogenic selenium nanoparticles. After an incubation time of 24 hours, the cell growth was analyzed using MTS assay.

Results:It is demonstrated that the bacterial strains used can generate biogenic selenium nanoparticles, concluding that selenium is a thermodynamically favorable metal ion that can be reduced by some reductase enzymes within the cells. TEM images of the nanoparticles showed nanoparticles with a constant distribution size surrounded by cellular debris. The chemistry of the samples was confirmed using EDX analysis, showing the distinct peaks of selenium. The antimicrobial activity of these biogenic selenium nanoparticles was tested against *Escherichia coli*and *Staphylococcus aureus*to discover their antibacterial ability. Nanoparticles can produce a decay in

the standard bacterial growth. Besides, the growth of *Staphylococcus aureus*was significantly inhibited for those nanoparticles synthesized by *Staphylococcus aureus*. Colony forming unit assays showed that a bigger concentration of selenium nanoparticle means a better inhibition of the bacterial growth. In vitro cytotoxicity assays were performed with human dermal fibroblasts (HDF) cells. The experiments showed that all nanoparticle concentrations tested did not inhibit the growth of cells while also confirming a high percentage of cell viability. Furthermore, after the incubation period, the growth of the cells was enhanced in comparison with the control without the nanoparticles.

Conclusions: Current and main methods to synthesize selenium nanoparticles use approaches which employ chemical methods. The weakness of these procedures (extreme reaction conditions, production of toxic byproducts, and need of purification) calls for a necessity of alternative approaches. Bacteria can be used to overcome these drawbacks naturally. Therefore, biogenic selenium nanoparticles were synthesized using different bacteria strains through an eco-friendly and green approach. Selenium nanoparticles with small and uniformly shape were synthesized with a relatively homogeneous size distribution. Decay in the bacterial growth was achieved against both Gram positive and Gram negative bacteria, showing no significant cytotoxicity effect when they were cultured with human dermal fibroblasts (HDF) cells.

538227 Nanoescapology Enabled By Surface-Engineered Magnetite: Novel Routes for Targeted Drug Delivery

Abstract:

Nanomedicine has become one of the most attractive avenues to promote targeted delivery and increase bioavailability of drugs that fail to spontaneously pass the cell membrane due to their high hydrophobicity and low solubility. As a result, over the past two decades an enormous body of knowledge has been developed where nanostructured materials served as the main platforms for conjugation and subsequent intracellular delivery of various pharmacological agents for the treatment of different conditions including several types of cancer, Alzheimer's and Parkinson's. Despite the important success both in vitro and vivo of the proposed vehicles, a major challenge is to assure that the delivered therapeutics remain active upon delivery. This is mostly due to the formation of endosomes that act as traps for the delivered molecules. To overcome this major issue, we proposed the conjugation of a transmembrane protein from *E.Coli* to nanostructured materials and particularly to magnetite nanoparticles, which were prepared by co-precipitation of Iron Chlorides. Immobilization was conducted after silanation, which rendered free amine groups on the nanoparticlesâ€™ surface. Protein molecules were then conjugated with the aid of glutaraldehyde to the N-terminal by forming imine bonds. As-synthesized and silanized magnetite size distribution in distilled water at 1 mg/mL was estimated via DLS. Immobilization was verified with the aid of FTIR analysis, SDS-PAGE and BCA assay. As-synthesized magnetite nanoparticles exhibited an average diameter of 120 Â±62.7 nm, which was maintained even after silanation. Successful protein conjugation was determined by the absorption bands of the silanation agent and the protein. Accordingly, vibrational bands at 3000 cm^{-1} and ~1600 cm^{-1} were identified and correlated to the terminal $-NH_2$ group of the silanation reagent APTES while the protein was correlated to the vibration of the amide I band at about 1650 cm^{-1}. The vehicle was successfully tested in liposomes, Vero cells and Monocytes. Briefly, fluorescently-labeled proteins were immobilized, delivered and imaged with the aid of a confocal microscope. Our findings suggest that the obtained nanobioconjugate is capable of translocating the cell membrane and bypassing intracellular endosomal routes. Additionally, citotocixity assays on Vero via LDH cells suggested a rather high cell viability for nanobioconjugate concentrations below 100 ug/mL. Due to the potency of this novel vehicle, we are exploring applications in the delivery of therapeutics for the treatment of Parkinson's disease in primary Astrocyte culture as well as anticancer compounds in a colon carcinoma cell line.

538453 Novel Glucosylceramide Synthase Inhibitor Based Prodrug Copolymer Micelles for Delivery of Doxorubicin

Abstract:

In order to improve the efficacy of chemotherapy for cancers, we have developed a novel prodrug micellar formulation to co-deliver ceramide-generating anticancer agents and ceramide degradation inhibitor (PPMP). The prodrug nanocarrier is based on a well-defined POEG-*b*-PPPMP diblock copolymer. The hydrophilic block of POEG-*b*-PPPMP is POEG, and the hydrophobic block is composed of a number of PPMP units, which could work synergistically with loaded anticancer drugs. POEG-*b*-PPPMP was readily synthesized via a one-step reversible addition-fragment transfer (RAFT) polymerization from a PPMP monomer. The newly synthesized polymers were self-assembled into micelles and served as a carrier for several hydrophobic anticancer drugs including DOX, PTX and ceramide. POEG-*b*-PPPMP prodrug polymer formed uniform micelles with a size of ~100 nm and exhibited intrinsic antitumor activity *in vitro* and *in vivo*. DOX-loaded POEG-*b*-PPPMP micelles exhibited an excellent stability of 42 days at 4 Â°C. In addition, POEG-*b*-PPPMP prodrug polymer was comparable to free PPMP in enhancing the production of ceramide as well as down-regulating the mRNA expression of GCS. Moreover, DOX loaded in POEG-*b*-PPPMP micelles showed higher levels of cytotoxicity than DOX loaded in a pharmacologically inert polymer (POEG-*b*-POM) and Doxil formulation in several tumor cell lines. Consistently, in a 4T1.2 murine breast cancer model, the tumor inhibition followed the order of DOX/POEG-*b*-PPPMP > DOX/POEG-*b*-POM â‰¥ Doxil > POEG-*b*-PPPMP> POEG-*b*-POM. Our data suggest that POEG-*b*-PPPMP micelles are a promising dual-functional carrier that warrants more studies in the future.

538858 Influence of Micro and Nanoscale Surface Roughness on the Wetting Characteristics of Flat Surface

Abstract:

Unique wetting characteristics of natural surfaces inspired researchers to study the wetting behavior of rough surfaces. In most of the artificial flat surfaces, the surface roughness is tailored to enhance solid-liquid adhesion, self-cleaning, anti-bacterial and directional controllability. The wettability (superhydrophobicity and superhydrophilicity) of a surface is mainly influenced by the geometric parameters of micro and nano-size pillared surface i.e. height, width and pitch of the pillar, and their relationship. In this work, the effect of varying height, width, and pitch of the pillars on the wetting state of the micro-patterned surface is investigated. Micro-structures are fabricated in a controlled manner on Silicon wafer (100) surface, with pillars height (50 µm, 70 µm, and 90 µm), width (30 µm, 45 µm, and 60 µm), and pitch (30 µm, 60 µm, and 90 µm) using photolithography. Scanning Electron Microscope (SEM) and Optical Profilometer confirm the different feature sizes on silicon wafers. These patterned surfaces are categorized into three types: 30 µm width, 40 µm pitch and height varies in the range 50 to 90 µm of the pillars are set as Type I, 70 µm height, 30 µm pitch and different width values from 30 to 60 µm of the pillars are set as Type II, and 70 µm height, 45 µm width and pitch varies in the range 30 to 90 µm are set as Type III. The wettability of the patterned surface is obtained from the contact angle data using sessile drop method. It is observed that increasing pillar height and decreasing the pitch of the pillars allow the droplet to rest in the

with strong solid-liquid adhesion at the interface. The apparent contact angle is strongly influenced by the contact angle hysteresis (CAH) caused by the feature size of the micro-structures. The transition in the wetting state is also studied with the droplet evolution during evaporation. Critical value of pillar height and pitch is determined to discuss the stability and transition of a droplet from Cassie Baxter to Wenzel state. The present work is also compared with the randomly designed micro/nano- scale surface roughness on the glass surface. The results demonstrated the extent and nature of surface engineering required to design a superhydrophobic surface.

542927 Synthesis and Characterization of Hollow Gold Nanoparticles for Gene Delivery

Abstract:

Intracellular drug and gene delivery requires overcoming the barrier to endosome escape. Genetic materials and protein function in the cell cytosol, but typically < 2% of mRNA delivered via synthetic vectors reach their destination. Our strategy is to use liposomes coated with gold nanoparticles that can be triggered by picosecond pulses of near-infrared light to release biomolecules from the endosomes. The NIR light pulses generate vapor nanobubbles that mechanically rupture the liposome and endosome, releasing the mRNA to the cell cytosol. Here we describe two methods of attaching gold nanoparticles to liposomes to generate nanobubbles. The first method creates plasmon resonant hollow gold nanoshells (HGN) via a galvanic replacement reaction of gold salts with cubic silver template nanoparticles. These HGN are tethered to the liposomes via thiol terminated polyethylene glycol lipids on the liposome surface. In the second method, plasmon resonant gold nanoparticles are directly condensed onto the liposome exterior via reduction of chloroauric acid. We examine the threshold fluence and wavelength dependence necessary to rupture the liposomes and release mRNA into solution to determine an optimal mRNA delivery system.

546357 Xeno Nucleic Acid Nanosensors for Enhanced Stability

Abstract:

The omnipresence of salts in biofluids creates a pervasive challenge in designing sensors suitable for *in vivo* applications. Fluctuations in ion concentrations have been shown to affect the sensitivity and selectivity of optical sensors based on single-walled carbon nanotubes wrapped with single-stranded DNA (ssDNA-SWCNTs). We herein observe fluorescence wavelength shifting for ssDNA-SWCNT-based optical sensors in the presence of divalent cations at concentrations above 3.5 mM. In contrast, no shifting was observed for concentrations up to 350 mM for sensors bioengineered with increased rigidity using xeno nucleic acids (XNAs). Transient fluorescence measurements reveal distinct optical transitions for ssDNA- and XNA-based wrappings during ion-induced conformation changes, with XNA-based sensors showing increased permanence in conformational and signal stability. This demonstration introduces synthetic biology as a complementary means for enhancing nanotube optoelectronic behaviour, unlocking previously unexplored possibilities for developing nano-bioengineered sensors with augmented capabilities.

546358 Restriction Enzyme Analysis of Double-Stranded DNA on Pristine Single-Walled Carbon Nanotubes

Abstract:

DNA has been extensively studied due to its versatility as a dispersant of SWCNTs. The specificity and binding affinity of DNA to various analytes has been shown to be strongly sequence dependent. As a result, through modifications of the nucleotide sequence, the DNA conformation on the SWCNT surface can be tailored to suit specific needs for techniques such as single-molecule detection, in vivoimaging, and chirality separation. To date, the majority of research has focused on the interaction between single-stranded DNA and SWCNT surface, with less focus on the nature of the interaction between double-stranded DNA (dsDNA) and SWCNTs. As a result, the exact interaction mechanism governing this type of complex remains strongly debated and largely unknown. In this study, we employ various biochemical methods to infer the conformation of dsDNA on the surface of SWCNTs. Our methods are based on imaging techniques for identifying DNA-modifying enzyme activity in the presence of dsDNA-SWCNT complexes. Our findings suggest that dsDNA can partially retain its native conformation on the SWCNT surface, and the degree of dsDNA accessibility is strongly sequence dependent. These findings offer new possibilities for SWCNT sensing applications that employ dsDNA, such as the optical detection of DNA-protein interactions

546359 Near-Infrared Confocal Imaging of Single-Walled Carbon Nanotube Uptake in Bacteria

Abstract:

The distinctive properties of single-walled carbon nanotubes (SWCNTs) have inspired the development of innovative applications in the field of cell nanobiotechnology. Most studies to date have focused on eukaryotic cells capable of internalizing SWCNTs functionalized with a variety of non-covalent wrappings. However, the effect of SWCNT functionalization on transport across the thick cell wall of prokaryotes remains unexplored. In this study, we explore the uptake of SWCNTs in Gram-negative cyanobacteria and demonstrate selective internalization of SWCNTs decorated with charged protein wrappings. The functionalized SWCNTs are shown to traverse the outer cell wall of both filamentous and unicellular strains of cyanobacteria, independent of whether the strain is naturally competent for DNA uptake, with adsorption and internalization rate constants of $k_{ads} = (9.08 \pm 0.16) \times 10^{-8} s^{-1}$ and $k_{in} = (1.466 \pm 0.011) \times 10^{-4} s^{-1}$, respectively. A custom-built, spinning disc confocal microscope was used for the first time to directly image near-infrared (NIR) SWCNT fluorescence within cells, revealing a highly inhomogeneous distribution of SWCNTs that is otherwise overlooked using conventional NIR widefield imaging. The nanobionic cells show sustained photosynthetic activity and growth, offering a powerful avenue for engineering photosynthetic organisms with augmented and even inherited nanobionic capabilities.

546360 Spinning-disc confocal microscopy in the second near-infrared window

Abstract:

Fluorescence microscopy in the second near-infrared optical window (NIR-II, 1000-1350 nm) has become a technique of choice for non-invasive *in vivo* imaging. The deep penetration of NIR light in living tissue, as well as minimal or complete absence of tissue autofluorescence within this range, offers increased resolution and contrast with even greater penetration depths. Here, we present a custom-built spinning-disc confocal laser microscope (SDCLM) that is specific to imaging in the NIR-II. The SDCLM achieves a lateral resolution of 0.5±0.1 μm and an axial resolution of 0.6±0.1 μm, showing a ~17% and ~45% enhancement in lateral and axial resolutions, respectively, compared to the corresponding wide-field configuration. We furthermore showcase several applications that demonstrate the use of the SDCLM for *in situ*, spatiotemporal tracking of NIR particles and bioanalytes within both synthetic and biological systems.

546361 Mediatorless, Reversible Optical Nanosensor Enabled through Enzymatic Pocket Doping

Abstract:

Single-walled carbon nanotubes (SWCNTs) exhibit intrinsic near-infrared fluorescence that benefits from indefinite photostability and tissue transparency, offering a promising basis for *in vivo* biosensing. Existing SWCNT optical sensors that rely on charge transfer for signal transduction often require exogenous mediators that compromise the stability and biocompatibility of the sensors. This study presents a reversible, mediatorless, near-infrared glucose sensor based on glucose oxidase-wrapped SWCNTs (GOx-SWCNTs). GOx-SWCNTs undergo a selective fluorescence increase in the presence of aldohexoses, with the strongest response toward glucose. When incorporated into a custom-built membrane device, the sensor demonstrates a monotonic increase in initial response rates with increasing glucose concentrations between 3×10^{-3} and 30×10^{-3} M and an apparent Michaelis–Menten constant of $K_{M(app)} \approx 13.9 \times 10^{-3}$M. A combination of fluorescence, absorption, and Raman spectroscopy measurements suggests a fluorescence enhancement mechanism based on localized enzymatic doping of SWCNT defect sites that does not rely on added mediators. Removal of glucose reverses the doping effects, resulting in full recovery of the fluorescence intensity. The cyclic addition and removal of glucose is shown to successively enhance and recover fluorescence, demonstrating reversibility that serves as a prerequisite for continuous glucose monitoring.

537352 Selectivity Enhancement of Nanowire Gas Sensors Using Impedance Spectroscopy and Artificial Neural Network

Abstract:

There is a significant market growth opportunity for sensors with better performances in the gas detection segment driven largely by more stringent safety regulations in monitoring and controlling industrial and residential environment. Because nanowires have ultra-high surface to volume ratio, organic nanowire sensors show high sensitivity, and fast response, coupled with low power consumption and low cost. However, their poor selectivity is a major challenge that prevents organic nanowire sensors from being efficiently used. In this talk, a new nanowire sensor/sensor array fabrication method, and enhancement of selectivity by using artificial neural network (ANN) models will be presented. Our previous and ongoing research show that copper tetracyanoquinodimethane (CuTCNQ) and tetrathiafulvalene charge transfer salts (TTF) can grow into nanowires from solution on photolithographic gold pattern in ambient temperature and pressure by simply applying overpotential. After drying in air, the CuTCNQ sensor can detect sub-ppm level ammonia gas by measuring the impedance spectroscopy. Equivalent electric circuit models will be used to analyze the conduction and detection mechanisms of the nanowires as well as the effect of their geometry and contact junctions. By employing the same fabrication method, different sensing materials can be deposited onto one devices working as a sensing array. The impedance data of different material in the sensing array will be coupled with ANN models to enhance the selectivity of the sensors. The ANN-enhanced sensors can selectively detect different gases, such as ammonia and ethanol in the ppm level.

Effects of CeO_2 in CuO-ZnO catalyst for the deep purification of CO derived from olefins at ambient temperature

HUANG Jinhua[1,2], YE Liping*[1,2], YANG Bingxing[1,2], Kong Meng [1,2]

(1. Shanghai Research Institute of Chemical Industry Co., Ltd, Shanghai 200062, China;

2. State Key Laboratory of Polyolefin and Catalysts, Shanghai 200062, China;

3. Shanghai Key Laboratory of Catalysis Technology for Polyolefins, Shanghai 200062, China)

ABSTRACT: Based on traditional CuO-ZnO catalyst, two types of additives of Zr and Ce were incorporated by co-precipitation method for the deep purification of CO derived from olefins. The structures of resultant catalysts were characterized with various techniques such as XRD, N_2 sorption, H_2-TPR, CO-TPR, XPS, TEM and DFT calculation. The results showed that the incorporation of CeO_2 into CuO-ZnO catalyst increased the specific surface area and pore volume obviously. From H_2-TPR and CO-TPR curves, it was obvious that Ce increased both the reducibility and CO adsorption capacity of catalyst, which were critical to the catalytic activity. XPS spectrum demonstrated that more active species such as Cu^+ and lattice oxygen were improved due to the CuO_x being introduced into the CeO_2 lattices. DFT calculation furtherly certified that the Cu nano-particles could gain electrons from the Ce elements. As a result, part of the Cu^{2+} were reduced to Cu^+ during the preparation progress. This might be the major reason for the excellent purification ability of CuO-ZnO-CeO_2 catalyst at room temperature, because the combination of Cu^+ with CO (0.43 eV) was much stronger than that of Cu^{2+} (0.15 eV). Under the conditions of 25 °C and 3.0 MPa, the concentration of CO from the liquid propylene could be purified from 10×10^{-6} deep to less than 10×10^{-9} with a stability test for 150 h.

Key words: CO; purification; olefins; Ce; Cu; DFT

Received date:

**Corresponding author. Tel: +86-021-52813238; Fax: 021-52813238; E-mail: ylp_by@126.com*

This work was supported by the Shanghai Natural Science Foundation (No.15ZR1421500)8.

INTRODUCTION

With the development of polyolefin technology, the amount of CO derived from olefins is required to be purified from several ppm deep to less than 30×10^{-9} at room temperature, which can help the polymerization catalyst (such as metallocene catalyst) to maintain a high activity[1,2]. Compared with rectification, CO catalytic oxidation is known to be an effective method for the CO deep purification in industry[3].

During the past years, much attention has been given to CO oxidation. In 1989, Hartua[4] reported that supported gold catalysts were highly active for CO oxidation at low temperature. But for its high cost and low storage, the noble metals is not available in industry and requires other substitute catalysts. Cu-based catalysts have been found good candidates later because of their low costs and high activities for CO oxidation[5-7]. Many studies have been carried out on Cu-based catalysts, such as CuO, CuO-MnO_2, CuO-ZnO, CuO-Cr_2O_3, CuO-ZrO_2 and CuO-CeO_2[8-12]. Among these, CuO-ZnO material was reported to be an efficient catalyst, and has already been used in industry widely[13]. But it can only be effective to decrease the concentration of CO in olefins to 0.1×10^{-6} at high temperature (>70°C)[14,15].

To improve the low-temperature's efficiency of CuO-ZnO catalyst, two types of additives of Zr and Ce were incorporated by co-precipitation method in this work. The structures of resultant catalysts were characterized with various techniques such as XRD, N_2 sorption, H_2-TPR, CO-TPR, XPS, TEM and DFT calculation, and the effect of CeO_2 was analyzed in detail. As a result, CuO-ZnO-CeO_2 catalyst has great development potential in the industrial application for CO purification derived from olefins.

1 Experimental

1.1 Catalysts preparation

Cu-based catalysts of CuO-ZnO-ZrO_2 (CZZ), CuO-ZnO-ZrO_2-CeO_2 (CZZC) and CuO-ZnO-CeO_2 (CZC) were prepared by the co-precipitation method according to the following procedure: (1) $Cu(NO_3)_2$, $Zn(NO_3)_2$, $Zr(NO_3)_4$ or $Ce(NO_3)_3$ were dissolved in the deionized water (1 mol/L) and stirred for 30 min at room temperature to form stock solution; (2) Precipitation agent $(NH_4)_2CO_3$ (1 mol/L) was added dropwise together with the solution (1) under vigorous stirring and then aged at 80 °C for 2 h; (3) The as-obtained precipitates were filtered and washed by deionized water until the final pH value reached 7.0; (4) Lastly, the precipitates were dried at 110 °C for 12 h and then calcined at 400 °C for 6 h with a ramping rate of 2 °C/min.

1.2 Catalysts characterization

X-ray diffraction (XRD) patterns were obtained on a Rigaku D/max-RA X-ray diffractometer operated at 40 kV and 100 mA with a scanning rate of 4°/min, using Cu K_α radiation (λ = 0.15406 nm), the corresponding XRD patterns were collected from 10° to 80° with a step of 0.02°. N_2 adsorption-desorption measurements were carried out on a Micromeritics ASAP-2020M apparatus at 77 K. Transmission electron microscope (TEM) characterization was performed on a JEOL JEM-2010F. H_2 temperature-programmed reduction (H_2-TPR) and CO temperature-programmed reduction (CO-TPR) were obtained on a

FineSorb-3010D, about 50 mg catalyst was pretreated at 300 °C for 1 h firstly, then the sample was heated at a rate of 10 °C/min in 10% H_2/Ar or 10% CO/Ar after it was cooled to room temperature to obtain the reduction curves. X-ray photoelectron spectra (XPS) was performed on a PHI 5000C ESCA System spectrometer (USA) equipped with Al K_{α} radiation (hv= 1486.6 eV) at 14 kV and 250 W. All binding energies (BE) were adjusted to the adventitious C1*s* peak at 284.6 eV. This reference gave BE values with an accuracy at ± 0.1 eV.

1.3 DFT calcuation

All DFT calculations were carried out using the Vienna ab initio simulation package (VASP) [16,17]. The core-valence electron interaction was treated by using the project-augmented wave (PAW) [18,19] method. The Perdew, Burke, and Ernzerhof (PBE) [20] functional within the generalized gradient approximation (GGA) was employed to evaluate the electronic exchange and correlation energy. Since the bulk CuO has an antiferromagnetic ground state, the spin polarized calculations were performed. The calculations were also conducted involving on-site Coulomb corrections (DFT+U) [21,22] to correct for the description of the electronic properties of copper and ceria, where U = 7 eV and 5 eV were applied to the Cu 3d [23,24] and Ce 4f states [25,26,27]. The Cu (3d, 4s), Ce (4f, 5s, 5p, 6d, 6s), C (2s, 2p), andO (2s, 2p) were treated as valence electrons. For CuO(111), we used a 2 × 2 supercell ($Cu_{96}O_{96}$), with a 400 eV cutoff energy and a Monkhorst-Pack [28] grid of 1 × 1 × 1 k-points because of the large size of the slab (~11 × 12 $Å^2$). The slab thickness was six layers (~12 Å), with a 15 Å vacuum gap. For the Ce-doped system, a single Cu atom on the surface was replaced by a Ce atom ($Cu_{95}CeO_{96}$). For all structural optimizations, the bottom two layers were fixed, while the other layers were allowed to relax until the atomic forces reached below 0.05 eV/Å.

To estimate the binding energies of CO, the following equation was used,

$$E_{bind}=-\left(E^{fix}_{slab+CO}-nE_{CO}-E_{slab}\right)/n \quad (1)$$

where $E^{fix}_{slab+CO}$ is the total energy of the system involving the fixed slab with the adsorbed CO, E_{CO} and E_{slab} are energies of the gas-phase CO molecule and the surface slabs, respectively, and n is the number of adsorbed CO molecules.

1.4 Activity test

The gas-solid phase CO oxidation reaction was carried out in a stainless-steel fixed-bed reactor in which 1.156 g sieved catalyst (40-60 mesh) was loaded. The reactant was consisted of 3.0% CO, 3.0% O_2 and nitrogen as the balanced gas. The reaction were conducted under the following reaction conditions,gas hourly space velocity was 2000 h^{-1}, the reaction temperature range was from 50 °C to 150 °C and the reaction pressure was atmospheric pressure. The compositions of the effluent gases were measured by an on-line gas chromatograph (Agilent GC-6890N) equipped with a thermal conductivity detector (TCD).

The liquid-solid phase CO purification activity test in which liquid propylene containing 10×10^{-6} CO was employed as a model compound of olefin,was carried out in a stainless-steel fixed-bed reactor loaded with 3.667 g sieved catalyst (40-60 mesh). The reaction conditions were 25 °C , 8.0 $h^{-1}$and 3.0 MPa. The products were analyzed by GC-9560-HG (Shanghai Huaai Co.) equipped with a PDD detector (America VACOL Company).

The CO conversion was calculated according to the following equation:

$$\text{CO conversion (\%)} = \frac{CO_{in}-CO_{out}}{CO_{in}} \times 100$$

Here, CO_{in} is the mass of CO in the inlet reactant, CO_{out} is the mass of CO in the outlet product.

2 Results and discussion

2.1 Characterization results

2.1.1 XRD

The XRD patterns were applied to observe the bulk structure of the catalysts. Figure 1 showed the XRD patterns of the catalysts. Diffraction peaks at 2θ of (32.5°, 35.5°, 38.7°, 48.7°, 61.5°,66.2°), 2θ of 28.5° and 2θ of 31.8° were corresponding to CuO phase (PDF-48-1548), CeO_2 phase (PDF-43-1002) and ZnO phase (PDF-43-1002), respectively [9,29].

The intensities of CuO and ZnO diffraction peaks were decreased with the addition of CeO_2, suggesting that CeO_2 improved the dispersion of CuO and ZnO. The CeO_2 lattice parameters of CZZC and CZC calculated from the XRD were 0.5404 nm and 0.5386 nm, respectively, which were smaller than that of pure CeO_2 (0.5411 nm). Since the Cu^{2+} ionic radius (0.072 nm) is smaller than Ce^{4+} ionic radius (0.092 nm), it was supposed that some Cu^{2+} ions might be incorporated into the CeO_2 lattice to form a possible Cu-O-Ce solid solution, which resulted in the contraction of CeO_2 lattice [7]. Compared with CZZC, the CeO_2 diffraction peak of CZC catalyst was 28.5° which was shifted to a lower angle, meanwhile, CZC gave a smaller CeO_2 lattice parameter (0.5386 nm), all those results suggested that Zr^{4+} could restrain Cu^{2+} entering the CeO_2 lattice. It is worth to mention that Zr^{4+} ionic radius (0.084 nm) is lager than Cu^{2+} ionic radius (0.072 nm) and smaller than Ce^{4+} ionic radius (0.092 nm), so it was referred that Zr^{4+} can compete with Cu^{2+} to enter the CeO_2 lattice.

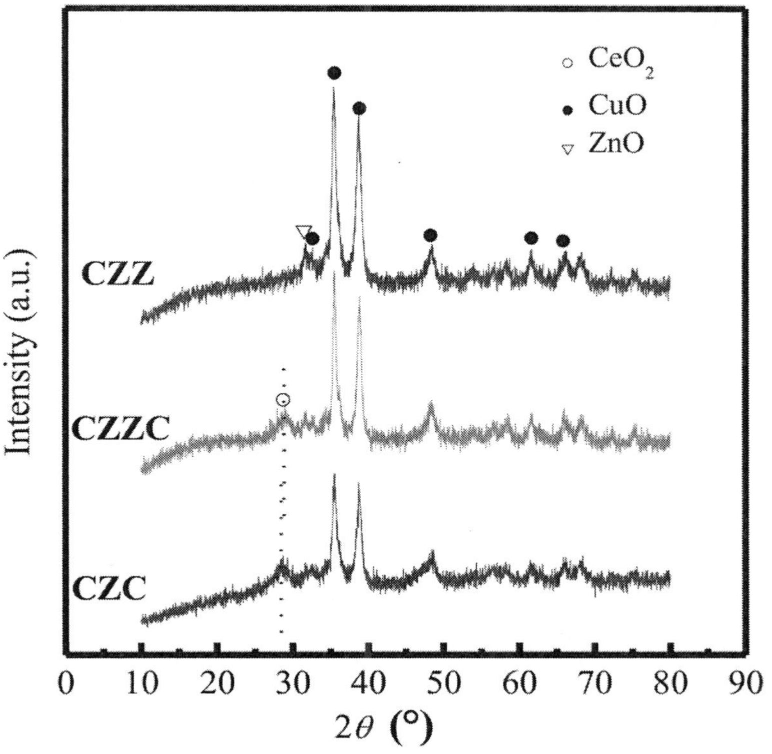

Fig.1 XRD patterns of catalysts

Table 1 showed the mean crystallite size of CuO particles in different catalysts which was calculated with Scherrer equation at 2θ of 35.5°. The crystallite size of CuO was increasing in the following order of CZC, CZZC and CZZ. CZC showed the smallest CuO crystallite size (5.8 nm), suggesting CuO dispersed well in CZC catalyst.

Table 1 Textural properties of catalysts

Sample	S_{BET} (m^2/g)	Pore Volume (cm^3/g)	Mean crystallite size of CuO (nm)[a]	Surface mass fraction(%)[b]			
				CuO	ZnO	ZrO_2	CeO_2
CZZ	66.83	0.484	9.50	63.6	22.1	14.3	-
CZZC	69.34	0.533	7.90	66.9	20.5	4.6	8.0
CZC	74.06	0.593	5.80	68.2	21.8	-	10.0

[a] calculated from XRD by Scherrer equation.

[b] calculated from XPS analysis.

2.1.2 N_2 sorption and TEM

Table 1 showed the textural properties of catalysts. The specific surface area and pore volume were placed in the increasing order of CZZ, CZZC and CZC. CZC showed the highest specific surface area (74.06 m^2/g) and the highest pore volume (0.593 cm^3/g). Combined with the result of XRD, it demonstrated that a higher specific surface area was in favor of the dispersion of CuO, thus CZC with a higher specific surface area and pore volume had a better dispersion of CuO and smaller CuO mean crystallite size.

Figure 2 exhibited the TEM images of the catalysts. A wormhole-like mesostructure which were aggregated by the metal-oxide nanoparticles was clearly observed for all samples. The particle size of CuO in CZC was about 20.4 nm, which was smaller than that in CZZ (26.3 nm) and CZZC (22.8 nm). The dispersion of CuO increased in the order of CZZ, CZZC and CZC. It suggested that CeO_2 improved the dispersion of CuO and decreased the CuO particle size, which agreed with the results of XRD and BET.

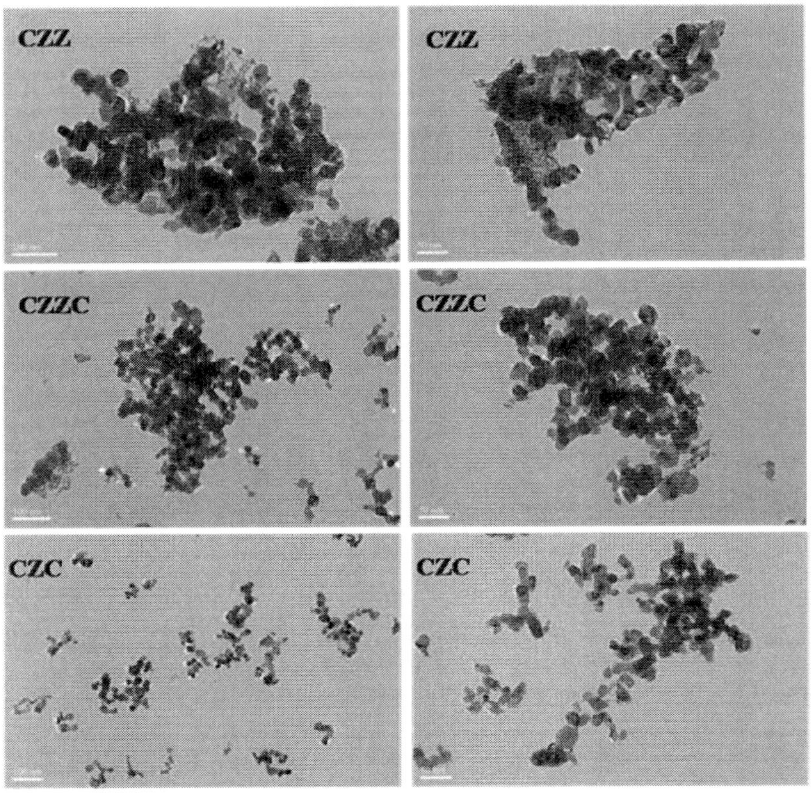

Fig.2 TEM images of the catalysts

2.1.3 H_2-TPR and CO-TPR

H_2-TPR and CO-TPR experiments were performed to investigate the reduction behaviors of the catalysts. Figure 3 and Figure 4 showed the curves of H_2-TPR and CO-TPR, respectively. As known, pure ZnO, ZrO_2 and CeO_2 cannot be reduced by H_2 or CO below 400 °C, thus all the reduction peaks were attributed to the reduction of CuO, which agreed with the results in literatures [30,31]. Compared with H_2-TPR, the reduction peak in CO-TPR was detected at a lower temperature (135-230 °C), suggesting that CuO was reduced easily in CO rather than in H_2.

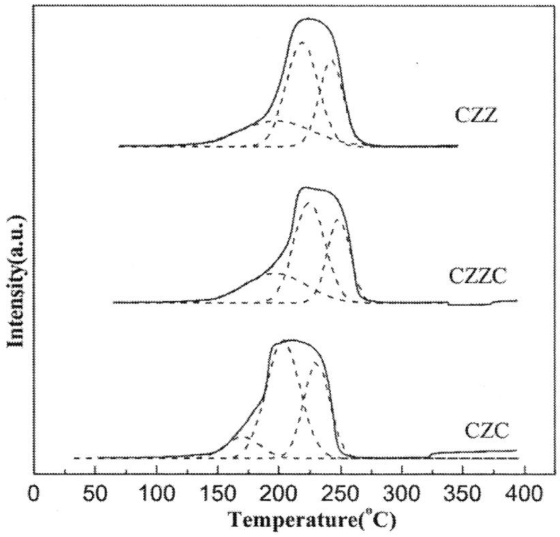

Fig.3 Profiles of H_2-TPR for catalysts

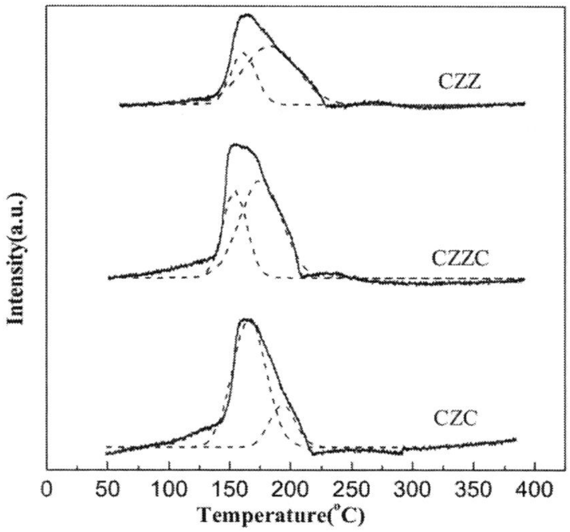

Fig.4 Profiles of CO-TPR for catalysts

Table 2 listed the Gaussian fitting results of H_2-TPR. Three reduction peaks were attributed to the reduction of CuO species in three different states [32,33]. The reduction temperature of the catalysts was in the decreasing order of CZZ, CZZC and CZC, and the relative H_2 consumption of Cu species increased slightly with the addition of CeO_2. Compared with CZZ, Ce modified Cu-based catalysts (especially CZC) showed a better reducibility and CO oxidation activity, indicating that CeO_2 promoted the reaction of Cu species in CO atomosphere. It was attributed to the improved CuO dispersion and the strong synergistic effect between Cu and Ce which were caused by the formation of Cu-O-Ce solid solution. In addition, Ce^{4+}/Ce^{3+} redox cycles and the electron conducting ability of CeO_2 were also contributed to the enhanced reducibility of the catalysts [34,35,36].

Table 2 Results of Gaussian fitting

Sample	Relative H_2 Consumption[a]	Relative H_2 consumption of H_2-TPR Gaussian peak[b] (Gaussian peak temperature/°C)		
		Peak 1	Peak 2	Peak 3
CZZ	1.00	0.260 (197.6)	0.457 (219.1)	0.283 (242.5)
CZZC	1.01	0.268 (196.7)	0.456 (224.8)	0.276 (248.5)
CZC	1.07	0.105 (170.1)	0.556 (203.4)	0.339 (230.6)

[a] The H_2 consumption amount of CZZ was designated as 1.00.

[b] The total H_2 consumption amount of H_2-TPR Gaussian peaks for CZZ or CZZC or CZC was designated as 1.00.

CO-TPR experiment was carried out to study the correlation between the reducibility of catalysts and the activity of CO oxidation [37,38]. Table 3 showed the Gaussian fitting results of CO-TPR. Different from the results of H_2-TPR, the reduction peak area of CZC increased significantly by 1.21 times than CZZ, the area of peak 1 increased by 2.23 times than CZZ simultaneously, it concluded that CeO_2 significantly increased CO adsorption capacity. However, the results of H_2-TPR indicated that the amount of reducible Cu species of CZC was only 1.07 times CZZ. Huang et al. [39] pointed out that the propensity behavior of Cu_2O toward valence variations in the oxygen-rich even oxygen-lean atmosphere at 140 °C or 205 °C made it seize or release surface lattice oxygen more readily, which led to the higher activity of Cu_2O than the other two copper species. Therefore, combined with Mar-van Krevelen reaction mechanism [40], it could be deduced that adsorbed CO can react with lattice oxygen preferentially even in lean oxygen atomosphere, which led to the notable increase of reduction peak area in CO-TPR. Additionally, CeO_2 improved the activity of lattice oxygen and promoted the reaction between absorbed CO and lattice oxygen, and further improved the activity of CO oxidation at a low temperature [41].

Table 3 Results of Gaussian fitting

Sample	Relative CO Consumption[a]	Relative CO consumption of CO-TPR Gaussian peak[b] (Gaussian peak temperature/°C)	
		Peak 1	Peak 2
CZZ	1.00	0.380 (166.2)	0.620 (193.4)
CZZC	1.13	0.386 (161.3)	0.614 (183.4)
CZC	1.21	0.700 (154.3)	0.300 (175.6)

[a] The CO consumption amount of CZZ was designated as 1.00.

[b] The total CO consumption amount of CO-TPR Gaussian peaks for CZZ or CZZC or CZC was designated as 1.00.

2.1.4 XPS

XPS was performed to further illuminate the surface composition and the chemical state of the elements existing in the catalysts. Table 1 listed the surface component mass fraction. The CuO content on the surface of the catalysts was in the increasing order of CZZ, CZZC and CZC. Combined with the results of XRD and H_2-TPR, it indicated that CeO_2 can not only increase the amount of Cu species but also the dispersion on the surface of the catalysts.

Figure 5 showed the O 1s XPS spectra of the catalysts. The spectra of the three samples showed two main peaks at 529.3 eV and 531.2 eV. According to the literatures [9,42,43], the peak at 529.3 eV can be ascribed to lattice oxygen in the metal oxides, while the peak at 531.2 eV can be assigned to adsorbed oxygen (such as H_2O or -OH). Compared with CZZ, the intensity of the peak at 529.3 eV increased obviously with the addition of CeO_2, it suggested that CeO_2 increased the amount of lattice oxygen on the catalyst surface, Yang [42] reported that CO reacting with the lattice oxygen gives a lower reaction barrier than that of adsorbed oxygen on Ce^{3+}, thus, it was reasonable that CZC catalyst with more lattice oxygen than adsorbed oxygen favored the better CO oxidation activity.

Figure 6 showed Cu 2p XPS spectra of the catalysts. According to the literatures [43,44], the peak at 934.1 eV with a shake-up satellite feature in the range of 940~950 eV is the characteristic of CuO, while the peak at 932.5 eV is ascribed to the reduced Cu species. The spectra suggested that Cu^{2+}, Cu^{+} and Cu were presented in the three catalyst., The peak area of the reduced Cu species increased with the addition of CeO_2, the proportion of the peak at 932.5 eV reflected by the ratio of $A_{932.5eV}/(A_{934.1eV}+A_{932.5eV})$ also increased obviously. It was attributed to the oxygen storage-release capacity and electron conducting ability of CeO_2 that promoted redox equilibrium between Ce^{4+}/Ce^{3+} and Cu^{2+}/Cu^{+} redox cycles ($Ce^{3+} + Cu^{2+} \leftrightarrow Cu^{+} + Ce^{4+}$), and then promoted the formation of Cu^{+} [9]. The spectra displayed that the reduced Cu species also existed in CZZ, it was possible that Zr^{4+} induced the cationic vacancies (in other words lattice defects) that resulted in the reduction of CuO and the formation of Cu^{+} [45,46]. However, CZC owned more Cu^{+} than CZZC, combined with the results of XRD, it may be explained that the existence of Zr^{4+} restrained the formation of Cu-O-Ce solid solution and weakened the synergistic effect between Cu and Ce.

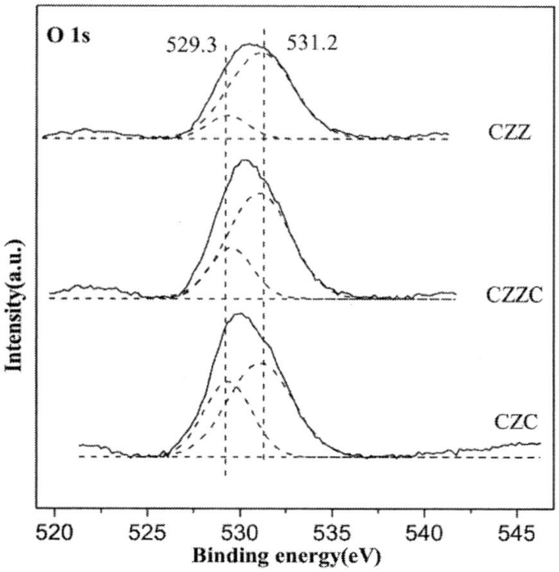

Fig.5 O 1s XPS spectra of catalysts

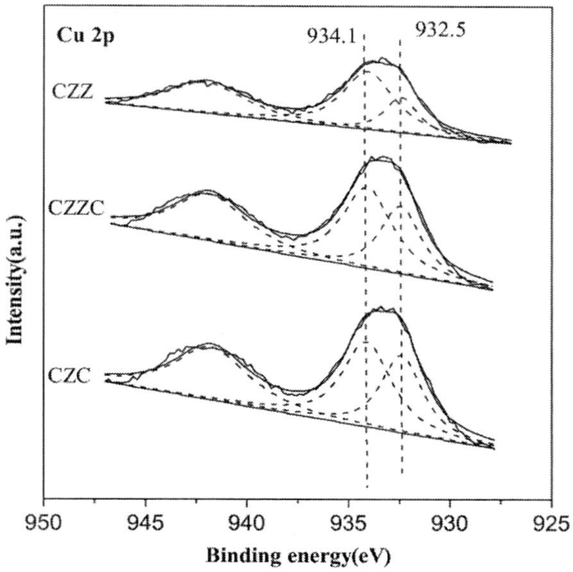

Fig.6 Cu 2p XPS spectra of catalyst

2.1.5 DFT Calculations

The calculated Ce-doped CuO(111) surface with one Cu atom replaced by a Ce atom was showed in Figure 7(a), together with the electronic structure (Figure 7(b)). It can be seen that, with the doping of Ce, an excess electron was localized on one Cu atom nearby. The Bader charge analysis showed that this Cu atom owned more charge (10.33 e) than other Cu atoms on the surface (about 9.98 e). Compared with the Bader charge of Cu^{+} (10.39 e) in bulk Cu_2O, this Cu ion should be considered as the Cu^{+}, while the other Cu ions were still Cu^{2+}. This result well demonstrated that CeO_2 promoted the formation of Cu^{+} for further.

To understand the effect of Cu^{+}, the binding energy (Table 4) of CO-Cu^{+} and CO-Cu^{2+} were calculated by using equation (1). The result showed that the combination of CO and the Cu^{+} (0.43 eV) was much stronger than that of Cu^{2+} (0.15 eV). In Figure 8, the optimized structures of CO bind with Cu^{+} and Cu^{2+} are presented. To further explain this discrepancy, we calculated the charge density difference and also performed the Bader charge analysis. As plotted in Figure 8 (a), CO molecule gains charge from the substrate at the Cu^{+} site, by as much as 0.18 electrons. While at the Cu^{2+} site, CO molecule as an electron donor gived 0.01 electrons to the surface (Figure 8(b)). The localization of excess electron on Cu^{+} caused more electron transfer between CO and surface, leading to the enhanced interaction between them. This result provided a powerful theory evidence for the importance of Cu^{+} in CO oxidation reaction.

Under the help of various characterization methods, such as FT-IR, DRIFTS and XANES, etc, it has been confirmed [39,47,48,49] that in CO oxidation, the propensity of Cu^{+} toward valence variations enables it to exhibit a better activity for the CO adsorption and then promotes CO oxidation reaction.

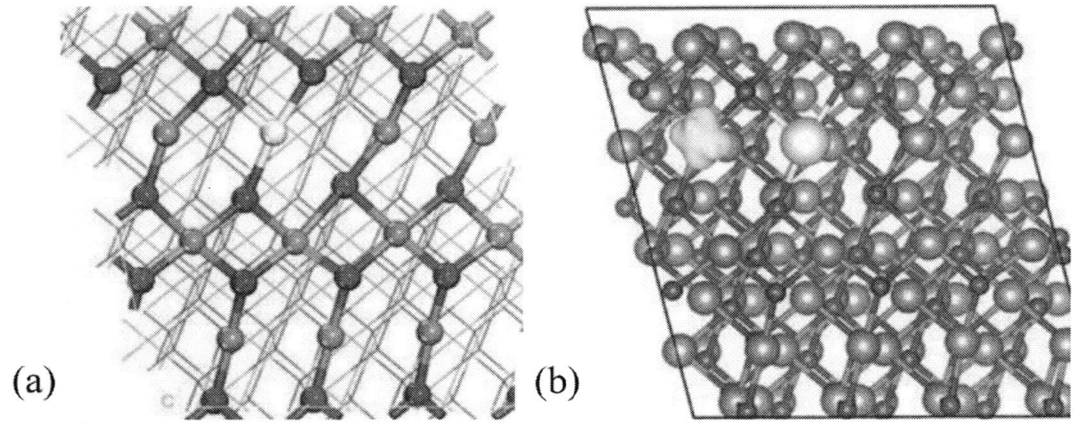

Fig.7 Top view of (a) calculated Ce-doped CuO(111) surface and (b) corresponding localized spin density structure. The Cu, Ce and O atoms are represented by balls in pink, red and white, respectively. The notation is used throughout this paper.

Table 4 Calculated binding energies and structural parameters of CO at different sites of CuO(111) surface.

Site	d(C-Cu)/Å	E_{bind}/eV
Cu^{+}	2.069	0.43
Cu^{2+}	2.121	0.15

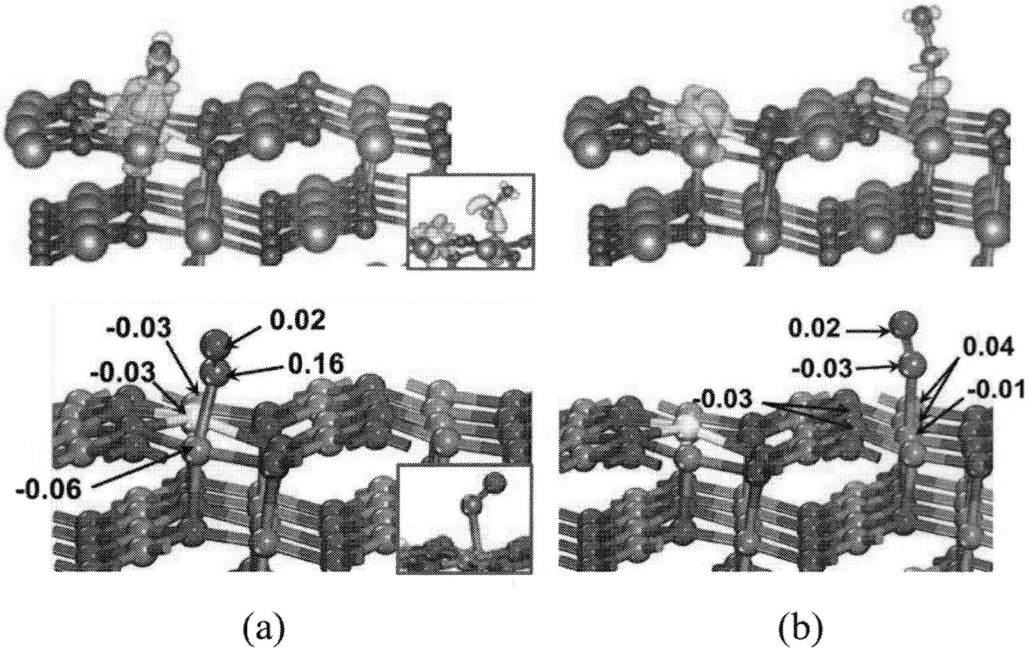

Fig.8 Isosurfaces (0.004 e/Å³) of charge redistribution (top panel) and Bader charges (bottom panel) for (a) CO/Cu⁺ and (b) CO/Cu²⁺. The yellow and blue isosurfaces denote charge gain and miss, respectively. Positive and negative signs of Bader charges are for gain and lose of electrons. The gray balls denote C atoms.

3 Catalytic performance of the catalysts for CO oxidation

The catalytic activities of the catalysts for CO oxidation as a function of reaction temperature were drawn in Figure 9. It was clear to observe that, the CO conversion increased significantly with the increase of reaction temperature. At the same reaction temperature, the CO conversion followed a decreasing order of CZC, CZZC, CZZ. The complete CO conversion temperature of CZZ, CZZC and CZC was pointed at 135 °C, 123 °C and 110 °C, respectively. It indicated that CO oxidation activity increased with the addition of CeO_2, and CZC showed the highest activity for CO oxidation.

Combined with the results of characterization and activity, it was concluded that CeO_2 promoter played an important role in CO oxidation reaction. CeO_2 increased specific surface area and pore volume, improved the reducibility and CO adsorption capacity of catalyst. In addition, CeO_2 promoted the formation of Cu-O-Ce solid solution and enhanced the synergistic effect between Cu and Ce, which increased the amount of Cu^+ and lattice oxygen, improving the CO oxidation activity at a low temperature.

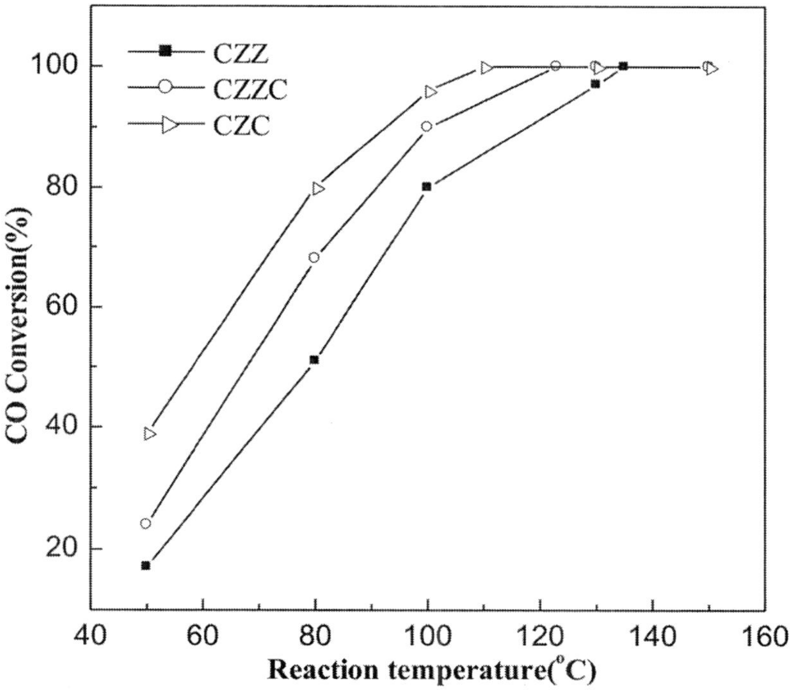

Fig.9 Catalytic Performance of catalysts for carbon monoxide oxidation in different temperature

4 Activity of catalyst in the CO deep purification derived from liquid propylene

Figure 10 showed the activity of $CuO\text{-}ZnO\text{-}CeO_2$ catalyst for CO deep purification derived from liquid propylene under the conditions of 25 °C and 3 MPa. It was evaluated that the concentration of CO can be purified from 10×10^{-6} to less than 10×10^{-9} with good stability for 150 h. Such results satisfied the requirement of the deep purification of CO for the polymerization reaction of olefins, providing a probable industrial application.

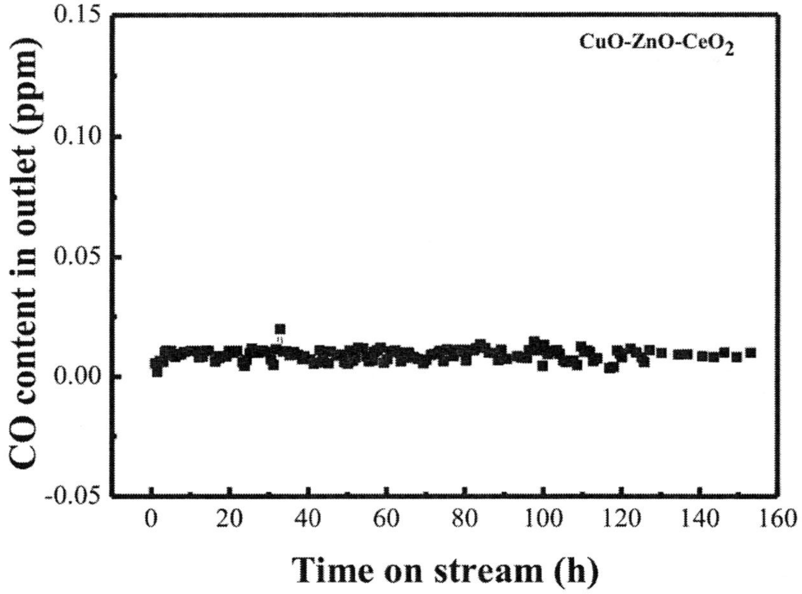

Fig.10 Activity of CuO-ZnO-CeO$_2$ catalyst for the CO deep purification derived from liquid propylene

5 Conclusion

There kinds of Cu catalysts were prepared and characterized for the deep purification of CO derived from olefins at ambient temperature. It was found that CeO_2 played an important role in the reaction of CO oxidation. The incorporation of CeO_2 into the CuO-ZnO increased the specific surface area and pore volume of the catalyst, Cu^{2+} ion can incorporate into the CeO_2 lattice and formed the Cu-O-Ce solid solution, which enhanced the synergistic effect between Cu and Ce. CeO_2 promoted the formation of Cu^{+} and lattice oxygen, therefore, CuO-ZnO-CeO_2 catalyst exhibited an excellent catalytic performance especially at a low temperature. Under the conditions of 25 °C and 3 MPa, the concentration of CO can be purified from 10×10^{-6} to less than 10×10^{-9} quietly in a long time test of 150 h, providing the catalyst has a probable industrial application for CO deep purification derived from olefins.

References

[1] S. Royer, D. Duprez. *Chem cat chem.*, 2011, 3(1): 24-65.
[2] J. R. Zhan, Y. Luo, L. P. Ye, et al. *Prog Chem.*, 2011, 30: 2424-2432.
[3] W. Liu, M. Flytzani-Stephanopoulos. *J. Catal.*, 1995, 153: 304-316.
[4] Haruta, N. Yamada, T. Kobayashi, et al. *J Catal.*, 1989, 115: 301-309.
[5] A. B. Lamb, W. C. Bray, J. C. Frazer. *J Ind Eng Chem.*, 1920, 12: 213-221.
[6] H. A. Jones, H. S. Taylor. *J Phys Chem.*, 1923, 27: 623-651.
[7] G. X. Chen, Q. L. Li, Y. C. Wei, W. P. Fang, et al. *Chin J Catal.*, 2013, 34: 322-329.
[8] M. Ricken, J. Noelting, I. Riess, et al. *J Solid State Chem.*, 1984, 54: 89-99.
[9] X. C. Zheng, S. H. Wu, S. P. Wang, et al. *Appl Catal A: Gen.*, 2005, 283: 217-223.

[10] X. L. Tang, B. C. Zhang, Y. Li, et al. *Catal Today.*, 2004, 93–95: 191-198.
[11] R. Pillai Unnikrishnan, Sarojini Deevi. *Appl Catal B: Environ.*, 2006, 65: 110-117.
[12] X. Y. Xi, S. Y. Ma, J. F. Chen, et al. *J Environ Chem Eng.*, 2014, 2: 1011-1017.
[13] J. Zhu, H. L. Li, C. Y. Liu, et al. China, 94115415.7[P]. 1996-03-06.
[14] Exxon Chemical Patents Inc. US, 9521146[P]. 1995-08-10.
[15] BASF SE (Ludwigshafen DE). EP, 1986772[P]. 2008-11-05.
[16] G. Kresse, J. Furthmüller. *Comput Mater Sci.,* 1996, 6:15-50.
[17] G. Kresse, J. Furthmüller. *Phys Rev B.,* 1996, 54: 11169-11186.
[18] G. Kresse, D. Joubert. *Phys Rev B.,* 1999, 59: 1758-1775.
[19] P. E. Blöchl. *Phys Rev B.,* 1994, 50: 17953-17979.
[20] J. P. Perdew, K. Burke, M. Ernzerhof. *Phys Rev Lett.,* 1996, 77: 3865-3868.
[21] S. L. Dudarev, G. A. Botton, S. Y. Savrasov, et al. *Phys Rev B.,* 1998, 57: 1505-1509.
[22] V. I. Anisimov, F. Aryasetiawan, A. I. J. Lichtenstein. *Phys Condens Matter.*, 1997, 9: 767-808.
[23] Y. Maimaiti, M. Nolan, S. D. Elliott. *Phys Chem Chem Phys.,* 2014, 16: 3036-3046.
[24] M. Nolan, S. D. Elliott. *Phys Chem Chem Phys.,* 2006, 8: 5350-5358.
[25] V. I. Anisimov, J. Zaanen, O. K. Andersen. *Phys Rev B.*, 1991, 44: 943-954.
[26] S. Fabris, S. Gironcoli, S. Baroni, et al. *Phys Rev B.,* 2005, 71: 041102-041105.
[27] M. Huang, S. Fabris. *J Phys Chem C.,* 2008, 112: 8643-8648.
[28] H. J. Monkhorst, J. D. Pack. *Phys Rev B.,* 1976, 13: 5188-5193.
[29] Y. Matsumura, H. Ishibe. *J Mol Catal A: Chem.*, 2011, 345: 44-53.
[30] I. Melián-Cabrera, M. L. Granados, J. L. G. Fierro. *J Catal.*, 2002, 210(2): 285-294.
[31] Elisa Moretti, Antonia Infantes Molina, Gabriele Sponchia, et al. *Appl Sur Sci.*, 2017, 403: 612-622.
[32] M. F. Luo, Y. P. Song, J. Q. Lu, et al. *J Phys Chem C.*, 2007, 111: 12686-12692.
[33] A. Martínez-Arias, D. Gamarra, M. Fernández-García, et al. *Catal Today.*, 2009, 143: 211-217.
[34] A. Martinez-Aiias, M. Femdez-Gareia, J. Soria, et al. *J Catal.*, 1999, 182: 367-377.
[35] W. Liu, A. F. Sarofim, M. Flytzani-stephanopoulos. *Chem Eng Sci.*, 1994, 49: 4871-4888.
[36] W. Liu, M. FlytZani-StephanoPoulos. *J Catal.*, 1995, 153: 317-332.
[37] D. H. Tsai, T. J. Huang. *Appl Catal A: Gen.*, 2002, 223: 1-9.
[38] J. B. Wang, D. H. Tsai, T. J. Huang. *J Catal.*, 2002, 208: 370-380.
[39] T. J. Huang, D. H. Tsai. *Catal Lett.*, 2003, 87(3–4): 173-178.
[40] Y. Liu, C. Wen, Y. Guo, et a1. *J Phys Chem C.*, 2010, 114(21): 9889-9897.
[41] B.X. Yang, Y. Luo, L.P. Ye. *J Mol Model.*, 2018, 24(20): 1-6.
[42] G. Avgouropoulos, T. loannides. *Appl Catal B.*, 2006, 67: 1-11.
[43] L. Qi, Q. Yu, Y. Dai, C. J. Tang, et al. *Appl Catal B: Environ.*, 2012, 119-120: 308-320.
[44] B. Wen, M. Y. He. *Appl Catal B: Environ.*, 2002, 37: 75-82.
[45] S. P. Huang, Y. C. Zhang, S. Y. Chen, et al. *Petro chem tech.*, 2009, 38(5): 482-485.
[46] H. Yun, H. Zhang, J. H. Chen, et al. *Acta Phys.-Chim.sin.*, 2004, 20(5): 524-528.
[47] D. Gamarra, C. Belver, M. Femandez-Garcia, et al. *J Am Chem Soc.*, 2007, 129(40): 12064-12065.
[48] H. Wan, Z. Wang, J. Zhu, et a1. *Appl Catal B: Environ.*, 2008, 79(3): 254-261.
[49] S. Hocevar, J. Batista, J. Lebec. *J Catal.*, 1999, 184: 39-48.

524862 Sensitizing Bacterial Cells to Antibiotics through Dynamic Topographyâ€"Triggered Biofilm Detachment

Abstract:

Microbial biofilm is a leading cause of chronic infections in humans and persistent biocorrosion and biofouling in industries due to extremely high-level tolerance of biofilm cells to antimicrobial agents. In the past three decades, the significance of biofilms has stimulated extensive studies on strategies for biofilm prevention and removal. However, due to the protection of extracellular matrix and slow growth of biofilm cells, controlling mature biofilm remains challenging. Recently, we reported that mature biofilms can be effectively removed (e.g. 99.9% detachment of 48h *Pseudomonas aeruginosa* PAO1 biofilms) using shape memory polymer-based dynamic change in surface topography. Here, we demonstrate that such biofilm detachment also sensitizes biofilm cells to conventional antibiotics. For example, concurrent treatment of 48h *P. aeruginosa* PAO1 biofilms during dynamic biofilm removal increased the susceptibility of released biofilm cells to tobramycin by more than 3 logs (2,479 fold) compared to surface-attached biofilm cells. The observed effects were attributed to the disruption of biofilm structure and increase in cellular activity as evidenced by a 11.8-fold increase in ATP level and 4.1-fold increase in expression of *rrnB* gene in detached cells. These results can guide the design of future combinational therapies for biofilm removal by integrating newly-designed smart dispersion strategies with currently available antibiotics.

536323 The Role of Flagellar Motor Reversals in Swarming in *Escherichia coli*

Abstract:

Swarming in thin fluid films on top of solid surfaces is enabled by the rotation of flagellar filaments in bacteria. To swarm, cells must overcome an increase in viscous drag due to the proximity to the surface. Previous reports suggest that cells likely compensate for the surface drag by promoting surface wetness through an unknown mechanism with the aid of a flagellar switch. The switch helps the filament erratically reverse the direction of rotation between clockwise (CW) and counterclockwise (CCW). The direction of rotation is dependent on the conformation of a switch-component, FliG. Here, we investigated the role of FliG conformation in facilitating bacterial swarming. We observed that swarming was abolished when FliG was locked in the CW conformation ($FliG^{CW}$), but not in the CCW conformation ($FliG^{CCW}$). Swimming was not abolished, so the conformation of FliG is likely critical for swarming. We found that torque generation in $FliG^{CW}$ strains was similar to WT strains. Swimming speed analysis revealed the loss of swarming ability correlated with a reduction in the flagellar thrust in the CW strain, which was likely due to a reduction in the number of filaments per cell. Fluorescence visualization showed that $FliG^{CW}$ swimmers have one less filament per cell on average compared to WT swimmers. Yet, $FliG^{CW}$ swarmers tend to lack filaments entirely (~57% of cells visualized carried no filaments, compared to ~10% in the WT). We hypothesized that there was a link between the function of the switch and degradation of the flagellar assembly; however, TIRF imaging revealed no significant degradation. Although the role of the switch in swarming remains unclear, our assays revealed that WT swarmer cells show a significant decrease in the probability of CW rotation (known as CW_{bias}) suggesting modifications are occurring at the level of the switch, and further pointing to its importance.

528605 Pseudomonas Aeruginosa Single-Cell Level Heterogeneity, Investigated Via Drop-Based Microfluidics

Abstract:

Single cell microbial studies can be challenging using traditional microbiology methods; however, single cell studies are critical in understanding cell-to-cell heterogeneity for interpreting bulk microbial growth behavior. This is especially true in biofilms, or microbial communities that express high levels of population heterogeneity and exceptional resilience due to its heterogeneity. Drop-based microfluidics, or the formation and manipulation of picoliter-volume drops of water-in-oil allows for the encapsulation of single cells, providing a platform for single cell studies. In this work, populations of Pseudomonas aeruginosa are examined at the single cell level within microfluidic drops incubated on a microfluidic chip over a period of 24+ hours. Confocal microscopy tracks the expressed plasmid gfp fluorescence of the cells. The resulting data is used to construct growth curves starting from single cells and to quantify the growth rate and lag phase. Our long-term growth study is performed using a device preparation technique that fills the porous microfluidic device with water-saturated carrier oil and allows for thermodynamic stability between the oil and aqueous phases. In absence of this preparation, drops are observed to shrink, foam, or evaporate from the device over time. Bulk studies of P. aeruginosa cells reintroduced to media after 4 days of starvation shows decreased growth in resuscitating cells. However, single cell studies using this incubation method show that these cells have increased and variable lag phase lengths, rather than decreased specific growth rates (Akiyama, et.al. PNAS, 2017). Our results demonstrate the quantification and monitoring of single cell behavior over 24+ hours, which will allow for a deeper understanding of the role of heterogeneity in bacterial populations.

537121 Differential Response of Mucoid and Non-Mucoid *Pseudomonas AeruginosaÂ* isolates to Interfacial Confinements

Abstract:

Fluid interfaces are energy-rich environments known to influence the self-assembly of small molecules or microparticles, and the formation of thin films by biological entities. The formation of bacterial films at fluid interfaces appears to be a dynamics process, in which cells coping with the existing interfacial energies respond to the entrapment by forming a matrix appropriate to their survival. We hypothesize that cells subjected to interfacial stress adapt and exhibit phenotypic changes essential to their survival. In this study, we are investigating the response to interfacial confinement of two clinical strains of *Pseudomonas aeruginosa* isolated from the airways of cystic fibrosis patients. To evaluate what phenotypic changes at fluid interfaces provide growth advantage, we are comparing a mucoid (PASL) and non-mucoid strain (PANT) of *P. aeruginosa* secreting various level of polysaccharides. We are investigating the mechanical properties of films formed by these cells to understand how they adapt and restructure environments with high interfacial energies. Some intrinsic properties of the cells confer viscoelastic behavior facilitating their adaption as characterized by the apparent moduli of the elastic films observed using pendant drop elastometry and particle tracking. Further characterization of whole genome and the transcriptomes of cells under interfacial confinements will provide new insights on the biological implications of interfacial films.

540115 Bloodmeal-Induced Inhibition of *Plasmodium* infection in Mosquito Vectors Using the Microbial Symbiont *Asaia*

Abstract:

Asaia sp. are gram-negative bacteria that have been shown to colonize the *Anopheles* mosquito midgut. Species of this mosquito are the vectors for malaria transmission throughout the world. Using a paratransgenic strategy in which transgenically modified symbiotic organisms affect their hostâ€™s phenotype, *Asaia* has been engineered to secrete anti-malarial effector molecules into the mosquito midgut, hindering its ability to carry *Plasmodium*. However, the constitutive expression of these effectors causes a fitness disadvantage to the transgenic *Asaia*.Therefore, it is desirable to express these molecules only when *Plasmodium*is present in the mosquito midgut, namely when a mosquito takes an infected blood-meal. Putative conditional promoters have been cloned into the plasmid pGLR1, which has a promoterless GFP reporter. These plasmids were then transformed into the lab strain *Asaia sp.SF2.1* and plated on minimal media enriched with blood. GFP fluorescent colonies were collected and screened both in-vitro, in liquid and solid media, as well as in-vivo, inside the mosquito midgut, to isolate the promoters that were only induced when blood meal conditions were present. Once the conditionality of these promoters was validated, they were cloned into a plasmid secreting the anti-malarial effector scorpine and transformed into *Asaia*. The increased fitness of these conditional strains was validated against a constitutively expressing control before testing each strain for *Plasmodium* inhibition. The anti-malarial strains were cultured in mosquitoes that were then fed on *Plasmodium*infected mice. The conditionally active promoter strains showed a significant reduction from the wild-type bacteria in the number of oocysts the pathogen formed as well as a significant reduction in disease prevalence effectively inhibiting the mosquitoesâ€™ ability to transmit malaria.

537630 Dynamics of Biofilm Elimination on Thermally Shocked Biomedical Surfaces

Abstract:

When bacteria colonize a medical implant surface, they form a biofilm which cannot be eradicated chemically. The current standard of care is surgical explantation of the device and surrounding tissue, with eventual reimplantation of a replacement device with twice the probability of infection. These infections are a $5 billion problem in the U.S. alone, impacting over 100,000 patients annually. One approach to mitigating these biofilms is *in situ* thermal eradication by generating a thermal shock directly at the device/biofilm interface. We have developed magnetic nanoparticle / polymer composite coatings which heat rapidly when exposed to an externally applied alternating magnetic field, and have quantified immediate bacterial population reductions of up to 6 orders of magnitude (i.e., "6-log" reductions).

Moreover, under certain circumstances the biofilm bacterial population continues to decrease for hours after the thermal shock is removed and physiological temperature is restored, resulting in complete eradication of the biofilm. Re-incubation studies at a variety of shock temperatures (50 – 80 °C) and exposure times (1 – 30 min) showed that thermal shocks producing immediate population reductions greater than 4-log resulted in complete elimination of the biofilm approximately 4 hours later, regardless of the temperature/exposure time combination used to achieve the immediate reduction. Biofilms subjected to thermal shocks with more modest immediate population reductions (less than 4-log) grew back to their initial population density within 24 hours. These investigations were extended to biofilms with different culture conditions producing pre-shock population densities 100 times greater than the earlier biofilms (nearly 10^9 CFU/cm^2). Despite the substantially larger initial population density, the same immediate population reduction (4-log) was required to prompt eventual die-off of the biofilm, even though the population density immediately after the thermal shock was 100 times greater than in the previous studies. This suggests that the degree of thermal shock required to eradicate a biofilm is independent of the biofilm's initial population density.

Furthermore, this approach may be integrated with other modifications to the biofilm/device interface to prevent or eliminate biofilms with even milder thermal shocks. Investigations of combined thermal shock and antibiotic exposure yielded population reductions whose orders of magnitude were greater than the sum of the orders of magnitude reduction for either treatment alone, enabling significantly milder thermal shocks to provide the same efficacy and providing an impetus for localized delivery at the device interface of antibiotics which are by themselves insufficient for biofilm elimination.

535582 Modelling Microbial Microenvironments through Encapsulation of Synthetic Communities

Abstract:

In the field of microbiology, it is believed that less than 1% of microbes are presently known and have been studied. The inability to culture many microbes using traditional methods, such as plating or suspension, has led to a vast gap in our understanding of microbiomes. To overcome this challenge, we have developed a novel microfluidics system which enables us to encapsulate microbes in a semipermeable membrane on the scale of nanolitres, eliminating resource competition and effectively increasing cell communication of isolated microcolonies.

In this study, we are introducing a mock community developed in the nanocultures with five microbial species, including *Pseudomonas aeruginosa, Escherichia coli Nissle 1917, Staphylococcus aureus, Streptococcus mutans and Candida albicans.* We determine how population density differs when cells are cultured in nanocultures as opposed to culturing them in suspension in a flask. After an incubation period of 24 hours, the nanocultures are ruptured and DNA is isolated and sequenced from viable cells. 16srRNA sequencing allows us to determine the abundance of each species in both the nanocultures and flask cultures, thus we can determine how the species interact with each other under the constraints of each system. We anticipate that our microfluidics device will be able to capture unculturable microbes in future applications, from a variety of different environments, such as oral, gut, marine and soil, allowing for unknown microbe discovery, and furthermore, a more accessible and in-depth study of these unknown microbes.

530520 Design and Engineering of Biohybrid Materials for Organic Electronics: From Supramolecular Assembly to Single Molecule Charge Transport

Abstract:

A grand challenge in the field of bioelectronics is to develop soft, deformable materials with hierarchical structures that possess a wide range of functionality. In order to address this challenge, new chemistries are critically required to generate robust bio-electronic interfaces between traditionally disparate materials. In this talk, I will discuss recent work in our group that focuses on biohybrid materials such as pi-conjugated oligopeptides that can be engineered for precise supramolecular assembly and charge transport applications. In the first part of the talk, I will describe the equilibrium and non-equilibrium assembly of pi-conjugated peptides using a combination of experiments and analytical modeling. In a series of experiments, pi-conjugated peptides are guided to assemble under reaction-dominated or diffusion-dominated conditions. Our results show that the morphology of self-assembled peptide fibers is controlled by the assembly kinetics. We further developed an analytical reaction-diffusion model to describe oligopeptide assembly, and experimental results are compared to the reaction-diffusion model across a range of parameters (Damkohler number). Moreover, we use microrheology to study the sol-gel transition of these materials during assembly, with direct measurement of modulus and photophysical properties during the sol-gel transition. *In situ* confocal fluorescence microscopy and *in situ* fluorescence lifetime imaging microscopy (FLIM) are also used to characterize peptides during the assembly process. In the second part of the talk, I will discuss the direct measurement of the electron transport properties and conductance of single molecules and oligomers. Here, we measure the conductance of single biomolecules and sequence-defined synthetic oligomers using a scanning tunneling microscope-break junction technique (STM-BJ). Using this approach, we characterize the electron transport properties of new classes of conjugated oligomers and biohybrid materials by varying primary monomer sequence. In this way, our work provides fundamental information underlying electron and charge transport behavior, which will be used to inform future design of molecular electronics materials.

525836 Lipid-like Materials for RNA Delivery: Predicting In Vivo Efficacy

Abstract:

Despite the promise of RNA therapeutics, progress towards the clinic has been slowed by the difficulty of delivering RNA drugs, such as short interfering RNA (siRNA) and messenger RNA (mRNA), into cellular targets within the body. RNA therapeutics are large (10^4 â€“ 10^6 g/mol) and negatively charged; they do not have favorable biodistribution properties in vivo nor an ability to cross the cellular membrane of target cells. In response to these challenges, our lab has created biodegradable, ionizable lipid-like materials called â€˜lipidoidsâ€™ that readily formulate into nanoparticles containing RNA. Lipidoids efficiently manipulate gene expression in a variety of biological systems, including hepatocytes, white blood cells and tumors. This talk will focus on the cell-free prediction of lipidoid efficacy in delivering mRNA to mice. Furthermore, I will describe a new formulation strategy for the synergistic co-delivery of mRNA and siRNA. Specifically, we found that siRNA-mediated gene silencing and mRNA-mediated protein expression improved 2-3 fold when they co-formulated in the same nanoparticle. Together, these data demonstrate the potential of lipidoid materials to robustly manipulate gene expression in vivo.

530980 Photoresponsive Nanomaterials in Tissue Repair and Radiotherapy

Abstract:

Electromagnetic radiation has a strong history of diverse applications in biotechnology and medicine. Plasmonic nanoparticles (e.g. gold or silver nanoparticles) can efficiently absorb and scatter incident radiation, which can engender transformative advances in biological imaging, diagnostics, and therapeutics. In this presentation, I will discuss our work on interfacing plasmonic nanoparticles with non-ionizing as well as ionizing radiation for applications in tissue repair and cancer radiotherapy. Plasmonic nanoparticles were incorporated within polypeptide matrices leading to the formation of laser-activated tissue-integrating sealant (LATIS) materials. Irradiation of these materials with non-ionizing, near infra-red light facilitated a photothermal response, which, in turn, resulted in rapid, fluid-tight sealing and repair of soft tissues both ex vivo and in live animals. Incorporation of plasmonic nanoparticles within polypeptide fibers led to the formation of laser-activated sutures which combine the benefits of conventional clinical practice (suturing) with photothermal sealing in a single device leading to much improved tissue repair in animals compared to that found with commercial sutures. Delivery of bioactive compounds along with laser sealing resulted in the acceleration of wound healing in mice. As a second application, I will describe our work radiation-induced formation of gold nanoparticles within gels as novel dosimeters for clinical radiotherapy. Irradiation of gold salt-loaded gels with electrons, photons (X-rays) or particles (protons) resulted in formation of gold nanoparticles, which was accompanied by a visible color change in the biomaterial. The intensity of the color change, induced by gold nanoparticle formation, was a function of the dose delivered. The gels were employed to determine and quantify complex dose profiles in 2D and 3D, which facilitates dose determination and planning in tissues and tissue models. The efficacy of the gel radiation dosimeter was evaluated using clinical anthropomorphic phantoms and in canine patients undergoing radiotherapy. Our studies demonstrate that plasmonic nanomaterials, in concert with ionizing and non-ionizing light, can facilitate novel advances in tissue sealing, wound repair, and clinical radiotherapy.

536282 Learning How to Predict SWCNT-Recognition DNA Sequences

Abstract:

In recent years, machine learning in bioinformatics has attracted considerable interest by its ability to transform large amounts of raw sequence data into useful scientific knowledge, without requiring explicit programming instructions. Many bioinformatics questions can be considered as examples of classification problems. In this work, we demonstrate the applicability of machine learning methods to discovery of special DNA sequences that recognize partner single walled carbon nanotubes (SWCNT). DNA/SWCNT hybrids have demonstrated significant potential in bio-applications because of their ability to disperse and sort SWCNTs by their chirality and handedness. Much work has been done to discover recognition sequences which recognize specific chirality of SWCNT, and significant progress has been made in understanding of the underlying structure and thermodynamics of these hybrids. Nevertheless, the success rate for *de novo* prediction of recognition sequences remains low. In this research, we investigate a new approach to predict recognition sequences using machine learning techniques and test the predicted results using aqueous two phase separation. Multiple sequence encoding methods (position-specific, term-frequency, combined or segmented term frequency vector, and motif-based feature) were used and compared. The transformed features were used to train several classifier algorithms (logistic regression, support vector machine and multilayer perceptron). The models were retrained each time by adding an experimentally tested new set of predicted sequences. Our predictive model showed significant improvement in ability to predict recognition sequences.

538744 Characterization of Double-Stranded DNA (dsDNA) on Single-Walled Carbon Nanotubes (SWCNTs)

Abstract:

DNA has been extensively studied due to its versatility as a dispersant of SWCNTs. The specificity and binding affinity of DNA to various analytes has been shown to be strongly sequence dependent. As a result, through modifications of the nucleotide sequence, the DNA conformation on the SWCNT surface can be tailored to suit specific needs for techniques such as single-molecule detection, *in vivo* imaging, and chirality separation. To date, the majority of research has focused on the interaction between single-stranded DNA and SWCNT surface, with less focus on the nature of the interaction between double-stranded DNA (dsDNA) and SWCNTs. As a result, the exact interaction mechanism governing this type of complex remains strongly debated and largely unknown.

In this study, we employ various biochemical methods to infer the conformation of dsDNA on the surface of SWCNTs. Our methods are based on imaging techniques for identifying DNA-modifying enzyme activity in the presence of dsDNA-SWCNT complexes. Our findings suggest that dsDNA can partially retain its native conformation on the SWCNT surface, and the degree of dsDNA accessibility is strongly sequence dependent. These findings offer new possibilities for SWCNT sensing applications that employ dsDNA, such as the optical detection of DNA-protein interactions.

528950 CNT-Based Carbon Monoxide Sensors with Voltage-Modulated Sensitivity

Abstract:

CNT-based Carbon Monoxide Sensors with Voltage-modulated Sensitivity

Suchol Savagatrup, Vera Schroeder, and Timothy M. Swager

Carbon monoxide (CO) outcompetes oxygen when binding to the iron center of hemeproteins, leading to a reduction in blood oxygen level and acute poisoning. Harvesting this strong and specific interaction between the CO and heme-inspired iron porphyrin provides promises for highly selective and customizable sensors that are viable alternatives to technologies based on mass spectroscopy and optical detection systems. In this presentation, we report the development of CNT-based sensors for the detection of CO with voltage-modulated sensitivity comprising iron porphyrin and functionalized single-walled CNTs, Figure 1. Modulation of the gate voltage led to that the *in situ* reduction of Fe^{3+} to Fe^{2+} enhances the interaction between the decorated CNTs and CO, leading to the significant increase in the signal strength when negative gate voltage is applied. UV-Vis, differential pulse voltammetry, and computational studies confirmed our sensing data. Furthermore, the sensors are selective to ppm levels of CO and functional in air. Our results provide a step toward the development of a novel platform of voltage-activated sensors based on highly specific chemical interactions.

538938 Self-Assembly of 3D Graphene/Carbon Nanotube Electrodes Via Electrostatic Polyanion Coordination for Biosensor Applications

Abstract:

Self-Assembly of 3D Graphene/Carbon Nanotube Electrodes Via Electrostatic Polyanion-Metal Nanoparticle Coordination for Biosensor Applications

Enoch A. Nagelli*, An B. Vu, Kamil Woronowicz, F. John Burpo, Alexander N. Mitropoulos
Department of Chemistry & Life Science
United States Military Academy, West Point, New York 10996
*Corresponding PI: Dr. Enoch Nagelli, Email: enoch.nagelli@usma.edu
The design, synthesis, and assembly of 2D graphene sheets with 1D carbon nanotubes (CNTs) can result in 3D nanostructures which can lead to the development of new biological/chemical sensors. Therefore, we developed a scalable aqueous solution-based process using poly(acrylic) acid (PAA)/CNT/graphene composite for advanced ultra-lightweight nanocomposites that are mechanically configurable and highly conductive. Using electrostatic self-assembly, we demonstrate an easy and scalable way to bind PAA/CNT composites with graphene oxide-coordinated Ni2+. The negative charge of many of the side chains of PAA in neutral pH water electrostatically bind with the Ni2+/graphene oxide to result in a polyelectrolyte-stabilized 3D carbon nanomaterial with a Ni2+ junction. This resulting 3D nanostructure is chemically reduced with sodium borohydride to form an all conductive polyelectrolyte/CNT-Ni nanoparticle â€“graphene composite. Therefore, our efforts are to contribute to the fundamental understanding of how specific carbon nanomaterials and metal nanoparticles and the combination of both can alter electronic band structure, doping, and affinity for other nanomaterials to be able to detect chemical and biological agents/analytes as well as transport hole or electrons through the 3D nanostructure.

539762 Fe_3O_4/graphene nanocomposites with Upsurge Superhydrophobic Properties

Abstract:

In earlier decade, graphene has revealed a copiousness of both fundamental science and potential applications due to its exceptional electrical, mechanical, thermal, and optical properties. Recently, increasing effort has been devoted to exploiting its new features, for example, wetting properties. Benefitting from its characteristic properties, graphene shows great potential for the fabrication of superhydrophobic surfaces, which could be potentially used for various anti-water applications like antifogging, self-cleaning, and anti-icing. In the present work, Fe_3O_4 nanoparticles fabricated graphene were cast-off for their wettability study. The composites show better superhydrophobic property than pure graphene and Fe_3O_4 nanoparticles. Inclusion of Fe_3O_4 nanoparticles to the graphene shows increase in water contact angle (WCA) from $105 \pm 1°$ to $159 \pm 1°$. Raman studies suggest this increase in water contact angle is due to increase in structural interaction between Fe_3O_4 nanoparticles and graphene. The composites were analyzed by different sophisticated analytical techniques. We further propose to examine Fe_3O_4/graphene composite antifogging ability in near future.

542620 The Impact of Shale Gas and Oil on the Chemical Industry

Abstract:

Lower alkanes and aromatics are very useful as both fuels and chemical raw materials. Occasionally supply disruptions constrained availability but also encouraged investigation of alternative feedstocks and chemistries. Recent new technologies for the production of unconventional gas and oil sources from tight shale formations significantly increased availability of both methane and condensate, radically altered process economics, resulted in the construction of much new intermediates capacity, and multiplied the ways traditional and alternative chemistries might be exploited. This presentation will outline historically how natural gas components became useful as chemical feedstocks, how government regulations occasionally led to perturbations in availability and price, how supply crises led to the development of alternative chemistries, catalysts, and ultimately the exploitation of shale resources. More specifically, will be discussed how excess coproducts, especially condensate, from increasing shale gas and oil production for energy are having significant unexpected impacts on the chemical process industries and opportunities for new process chemistry research.

541186 Sustainable Energy and Chemicals: Past, Present, and Future

Abstract:

The last 40 years has seen the U.S. energy situation vacillate from gasoline rationing and concerns over Peak Oil, to projected energy and feedstock prices that were noncompetitive in global markets for chemicals production, to the economic revolution of frack gas and light tight oil. The importance of low cost energy and feedstocks in driving economic opportunity will be discussed relative to this history, along with the importance of sustainability in decision making to gain stakeholder buy-in and acceptance. Strategies for managing career paths and competencies despite changing targets and high uncertainty will be addressed.

542618 Disruptions: What the Future May Hold

Abstract:

Disruptive forces are all around us. They change the way we communicate and work. They change the way we develop goods and services. They also change the types of products that are manufactured. Examples such as the way in which we produce and consume energy or the feedstocks we use to produce product have seen many changes over the course of the last few decades, and these disruptions are expected to continue. The chemical engineering discipline and the expectations for a chemical engineerâ€™s skills has also changed and adapted as a result of these disruptive forces. A discussion of the broad themes associated with these forces and the implications that they may have to chemical engineers in the future will be discussed.

542621 Geopolitical Factors Influencing the Evolution of the Chemical Industry

Abstract:

The basic chemicals and plastics industry is built on a simple principle: find a cheap feedstock and convert it into something more valuable. However, the right feedstock depends on place and time. Certain policies and actions of various governments have influenced the evolution and profitability of the chemical industry during the past 50 years, sometimes in surprising ways. For example, high oil prices and US government funding encouraged the development and commercialization of coal gasification technology during the late 1970s and early 1980s. This enabled China to develop a competitive basic chemical industry using coal as a feedstock during the last 25 years. Similarly, an abrupt increase in oil prices during the first decade of the 21st century encouraged the present shale gas revolution, causing a major shift in the economy of producing basic chemicals and polymers in the US. If history is our guide, then what governments choose to fund and regulate may be as important as market forces in determining the shape of our industry and profession in the future.

542622 Agility & Resilience: How to Maintain Career Competitiveness in the Changing Chemical Industry

Abstract:

The world around is changing. The shift to protectionism, global instability and price volatility of raw materials and feedstocks has the potential to wreak havoc on the chemical industry. With the revolution of shale gas in North America and the sudden drop in feedstock costs, the chemical sector in the US has reasserted its historic dominance. Learn how the role of a chemical engineer can contribute to industry growth and resilience, while adapting to allow for personal agility throughout the career life cycle.

540922 Mechanisms Contributing to the Formation of "Floating Biofilms" in Staphylococcus Aureus Orthopedic Infections (Invited Talk)

Abstract:

Despite the wide use of antibiotics, breakthrough infections frequently occur after orthopedic surgery. The major pathogen involved with such infections is Staphylococcus aureus. Here, we investigated the mechanisms that contribute to the exceptional recalcitrance of S. aureus to antibiotic treatment during joint infections by an ex-vivo approach using human synovial fluid. We found that even with antibiotic concentrations that far exceed the expected bactericidal levels, S. aureus is not eradicated in synovial fluid (SF).We show that this is due to the fact that in SF, S. aureus forms exceptionally strong biofilm-like aggregates that significantly increase resistance to antibiotic treatment. Screening a transposon bank identified S. aureus fibronectin- and fibrinogen-binding proteins as important for the formation of those macroscopic clumps, suggesting an important role of fibrin-containing clots in the formation of bacterial aggregates during joint infection. While these S. aureus surface proteins are a prerequisite for agglomeration, low activity of the Agr quorum-sensing regulatory system and resulting low production of the phenol-soluble modulin (PSM) surfactant peptides, which function by disrupting interactions of biofilm matrix molecules with the bacterial cell surface, cause agglomerates to grow to exceptional dimensions. Together, our findings support a two-step model of staphylococcal prosthetic joint infection: Interaction of S. aureus surface proteins with host matrix proteins such as fibrin initiates agglomeration; thereafter, the bacterial agglomerates grow to extremely large sizes owing to the specific conditions present in joints. Our findings provide a mechanistic explanation for the reported extreme resistance of joint infection to antibiotic treatment and have important implications for anti-staphylococcal therapeutic strategies.

540853 Bacteria Adhesion Is Mechanosensitive to Polymer Coating Properties (Invited Talk)

Abstract:

Biofilm-related infections are a class of notoriously difficult to treat healthcare-associated infections, which commonly develop on the surface of implanted medical devices. Despite lacking visual, auditory, and olfactory perception, bacteria detect and settle on surfaces, however, how the intrinsic properties of materials affect the initial adhesion microorganisms remains relatively unknown. By unveiling the structure-property relationships between polymer materials and microbial adhesion, we could guide the design of materials *a priori*to resist the adhesion of infection causing microorganisms, such as *Staphylococcus aureus*. In this presentation, I will discuss the effect that fundamental properties of polymer coatings (i.e., molecular architecture, stiffness, and thickness) have on the surface-associated transport of bacteria and on the adhesion of bacteria under quiescent conditions. By decoupling the effects of molecular architecture, stiffness, and thickness from coating chemistry, we have unlocked specific structure-property relationships that can be tailored to control the degree of bacterial adhesion and subsequently, reduce the formation of biofilms.

540784 Invited Talk 3:Â Prospective Technologies Targeting Microbial Biofilm and Its Microenvironment

Abstract:

Biofilm formation is a key factor for a wide range of microorganisms that cause many infections in humans and biofouling in industrial settings. Biofilms are highly structured microbial communities embedded in an extracellular matrix, a multifunctional scaffold that provides adhesion, cohesion and protection to the resident microorganisms. The matrix also provides a diffusion limiting milieu that facilitates the creation of localized chemical and pH microenvironments. These properties contribute to biofilm recalcitrance to mechanical removal and antimicrobial treatment, posing significant challenges for the development of effective antibiofilm therapeutics to control biofilm-associated diseases. This talk will focus on recent approaches developed in our laboratory to disrupt pathogenic biofilms using nano and biotechnology. Using in vitro, ex vivo and in vivo biofilm models with biophysical and super-resolution imaging methods, we have demonstrated how pH-activated nanoparticles or antimicrobials with matrix-degrading activity can target the complex biofilm microenvironment and help prevent biofilm-associated oral diseases. We will also discuss current challenges and future directions for the development of biofilm-targeting therapeutics.

526680 Investigating the Interfacial and Metabolic Properties of Bacteria at Hexadecane-Water Interfaces

Abstract:

Hydrocarbonoclastic bacteria interact with oil-water interfaces in a manner that continues to be unveiled. To degrade hydrocarbons, cells must overcome the challenges of the energies existing at the oil-water interface. The cells become trapped between the two fluid phases that exert interfacial tensions and shear stresses on the cells. We hypothesize that cells develop a protective interfacial layer owing to their ability to metabolize one of the phases. We anticipate that primary oil-eating microbes would exhibit improved intrinsic properties enabling them to withstand the interfacial stress compared to non-indigenous species. Here, we examine the interfacial behavior of *Alcanivorax borkumensis*, a primary oil-eating marine isolate. Using pendant drop elastometry, we characterize the mechanical properties of films of bacteria at the hexadecane-water interfaces (FBI) to understand how cells alter the physicochemical properties of the interfaces. Our results demonstrate that *A. borkumensis* develops a film with viscoelastic properties characterized by frequency dependent storage and loss moduli. We further characterize the interfacial dynamics of *A. borkumensis* under chemical stress to comprehend how cell activity affects the mechanical properties of FBI. Our results demonstrate that cells responding to chemical stress exhibit increased tendencies to alter the interface with time. Furthermore, we correlate data in the pendant drop mode to flat interfaces with particle tracking and microrheology techniques. This two-system approach elucidates the biological implications interfacial bacterial films in a manner relevant to bioremediation.

528984 Complex Liquid EmulsionsÂ and on-Chip Ring ResonatorsÂ for Bacteria Detection

Abstract:

Complex Liquid Emulsions and On-Chip Ring Resonators for Bacteria Detection

Suchol Savagatrup and Timothy M. Swager

Complex liquid emulsions are a class of sensors that translate chemical events into observable morphological responses. Recently, we reported selective and sensitive methods to detect the activity of enzymes and the presence of foodborne pathogens. These methods of transduction rely solely on the changes in the optical transmission of the emulsions upon changes in the morphology, which may limit the generality of the analyte. In this presentation, we report the novel approach to continuously sense and quantify the morphology of these emulsions using on-chip ring resonators. The morphological changes of the droplets are visualized by the characteristic of the transmission spectra of the resonators. Specifically, the shift in the resonant wavelength of the ring resonator is highly sensitive to even the small changes in the localized effective refractive indices induced by the morphological switches. By tuning the compositions of the droplets, we demonstrated the on-chip device capability and sensitivity to detect different classes of analytes. We anticipate the combination of complex liquid emulsions and integrated photonics will provide ultra-compact, cost-effective, and generalized method for other chemical and biological sensing schemes.

535514 Native Airway Mucus Rheology in Health and Patients with Cystic Fibrosis Having Positive or Negative Microbial Culture

Abstract:

In health, airway mucus is composed primarily of long, polymeric mucin proteins as well as some globular proteins and cell debris. Airway dehydration in cystic fibrosis (CF) gives rise to hyperconcentrated, viscoelastic mucus that is difficult for the mucociliary escalator to clear. This unclearable mucus traps pathogens in the lungs leading to inflammation and airway destruction. Because disease onset occurs first in the distal airways, sampling of airway mucus from infant and adolescent patient lungs often relies on the collection of Bronchioalveolar Lavage Fluid (BALF). While the lavage inherently dilutes the airway surface milieu, we show that roughly half (53%) of the mucus-forming mucin biopolymers in CF are present in the airways in insoluble "flakes" or "rafts" that are observable via microscopy. Because such a proportion of airway mucins are insoluble, we hypothesize that the characterization of these flakes will yield important advances in understanding the etiology of CF lung disease. Flakes in BALF present a number of research obstacles mostly arising from the difficulties associated with characterizing the biophysics of heterogeneous samples. To address this, we employ a combination of particle tracking microrheology (PTMR) and Gaussian Mixture Modeling (GMM) in order to isolate and study the biophysical properties and biomedical implications of these flakes.

We collected BALF from 29 patients with CF (age 10.5±4.4 years) and 6 healthy adults. We added 1μm fluorescent polystyrene beads to the samples to probe flake rheology via PTMR. The local complex viscosity (η^*, Pa·s) of BALF samples was computed from mean squared displacement of each individual bead and the generalized Stokes-Einstein equation. Bulk rheology was assessed in the CF samples using a cone and plate rheometer. Samples were scanned for flakes and videos of bead thermal motion in and near the mucus flakes were recorded at 60fps for 30s. Rheological signals were clustered into "watery" and mucus components via GMM.

Comparison between η^* measurements from bulk rheology and the ensemble mean of all beads in each CF patient sample revealed a moderate correlation ($r=0.42$, $p<0.05$). However, comparison of only the mucus component η^* from PTMR enhanced the correlation ($r=0.60$, $p<0.01$), indicating that bulk rheology is more reflective of the thickest components of a heterogeneous sample. Comparing CF samples to healthy adults, weighted η^* (i.e. the sum of the mean η^* in the water and mucus components multiplied by their mixing proportion, p_m) was increased in CF ($\log_{10}(\eta^*)=-0.96$) vs. non-CF (-2.5, $p<0.01$), but the mean of the mucus component was similar. The span of η^* in the mucus component was nearly 3 logs larger in CF samples, and mucin concentration was increased 3,000-fold in CF BALF relative to control samples. Furthermore, the GMM mixing proportion (p_m) of mucus was increased in CF ($p_m=0.83$) vs. non-CF ($p_m=0.27$) samples ($p<0.01$). Interestingly, neither hyperconcentration nor hypersecretion was observed in the subset (n=5) of patients with CF whose BALF was negative for microbial culture. Culture negative samples had weighted $\log_{10}(\eta^*) = -2.8$ and $p_m = 0.23$. The culture positive values of weighted $\log_{10}(\eta^*) = -0.75$ and $p_m = 0.85$ were significantly increased versus the healthy controls ($p<0.01$, both) and the culture negative patients with CF ($p<0.01$, both).

Through a combination of PTMR and Gaussian Mixture Modeling we find that while mucus flakes arise in health and in CF, hyperconcentration only occurs in CF mucus flakes, and orders of magnitude more mucus is found in CF samples. Importantly, increased rheological signal is only found in CF samples that were positive for microbial culture, indicating that infection may be linked to hyperconcentration and hypersecretion.

530457 Engineering Biology to Make Novel Antimicrobials

Abstract:

Proteins perform the cellular tasks required for life. The great variety of their biological activity is due in part to their vast combinatorial space: 20^n, n being the number of amino acids present in any given peptide chain and 20 being the number of natural amino acid monomer building blocks. Yet we do not have the tools to properly engineer these diverse molecules. One approach is to start small: I will present foundational synthetic biology frameworks to rationally develop peptides, tiny proteins that display great sequence diversity but are more amenable than larger molecules to redesign and engineering. My approach is to expand natureâ€™s repertoire to build novel synthetic peptides with extremely useful properties. My overarching vision is to generate a peptide encyclopedia encompassing peptides that target every medically relevant microbe and to devise therapies that nature has not previously discovered. The synthetic peptides that I am developing offer solutions to some of the most pressing unmet clinical challenges we face, including finding strategies for treating antibiotic-resistant infections.

537919 Association with Outer Membrane Vesicles Drastically Alters Bacterial Toxin Activity

Abstract:

Historically, most bacterial protein toxins have been purified and studied in their water-soluble, free form. The mechanisms of these toxins and their roles in disease have been well-established. However, it has recently been observed that many bacterial toxins can also be released in association with outer membrane vesicles (OMVs), which are produced from the outer membrane of Gram negative bacteria. Our lab has studied the activity of two bacterial toxins that are associated with OMVs and found significant differences in the delivery and activity of both when they are associated with OMVs compared to when they are free. *Vibrio cholerae*, the causative agent of cholera, secretes cholera toxin (CT), which binds to the ganglioside GM1 on host cells to initiate its toxic mechanism. We have discovered that most of the secreted CT is actually associated with OMVs, not in its free form, and in its OMV-associated form, the toxin does not bind to GM1. This finding highlights the need to understand delivery of this toxin in its OMV-associated form, as inhibitors of GM1 binding by CT are unlikely to be effective anti-virulence strategies. Similarly, the oral bacterium, *Aggregatibacter actinomycetemcomitans*, secretes a leukotoxin (LtxA) in a water-soluble form. LtxA recognizes cholesterol and an integrin receptor, LFA-1, on host cells. Because LFA-1 is only expressed by human immune cells, this toxin (in its water-soluble form) has been reported to be active only against these specific cells. However, OMV-associated LtxA is delivered to cells in an LFA-1-independent manner and is active in cells that do not express LFA-1. Importantly, this finding demonstrates that the activity of this toxin is much greater than has been suggested in the past. Together, our results indicate that bacteria package protein toxins in vesicles to enable long-distance delivery and to expand the range of activity of the toxin. In order to fully characterize the virulence of pathogenic organisms, it is essential to study these proteins in their vesicle-associated form, as well as their free form.

535937 Coal to Carbon Fiber

Abstract:

Aaron Owen[1], David Eaton[1], Rodney Andrews[1], Eric Eddings[2], and Matthew Weisenberger[1]

[1]University of Kentucky, Center for Applied Energy Research

Lexington, KY 40511

[2]University of Utah, Department of Chemical Engineering

Salt Lake City, UT 84112

Coal to Carbon Fiber

Demand for carbon fiber continues to grow at approximately 10% compound annual growth rate (CAGR). Over 90% of it is derived from polyacrylonitrile (PAN) precursor fiber, as it offers a high toughness carbon fiber for structural applications. Yet pitch-precursor-derived carbon fiber offers some key advantages over incumbent PAN-based carbon fiber. First, pitch can be melt spun, whereas PAN is solution spun. Melt spinning is significantly more economical and quite a bit faster than solution spinning; solution spinning is particularly expensive in terms of solvent use and recovery. Second, pitch typically has superior carbon yield, up to 80 wt.% conversion, while PAN offers approximately 50% conversion. There is future interest in the utilization of pitch for low-cost carbon fiber. Precursor pitches can be derived from petroleum and/or coal feedstocks. Coal tar is primarily generated from the overheads of coking processing of metallurgical coal. Upgrading coal tar to a pitch offers a viable path to high aromaticity pitches â€“ from which liquid crystal, or mesophase, pitch can be derived. Mesophase-pitch-based carbon fibers offer extremely high stiffness and thermal conductivity.

Here we report on our work utilizing coal tar pitch as a precursor for carbon fiber. We will discuss the coal tar extraction, and pitch upgrading processes. Quantifiable aspects of the pitch which are desired for high spinnability will be discussed. The melt spinning process will be presented, followed by general oxidation and carbonization processing. The resultant carbon fiber properties and yields will be discussed in the context of existing carbon fibers. Aspects of the overall value proposition of coal to carbon fiber will also be addressed.

CARBON NANOFIBERS FROM A BLEND OF LIGNIN WITH RECYCLED PET: PROPERTIES AND CHARACTERIZATION

Efstratios Svinterikos[1], *Ioannis Zuburtikudis*[2*], *Mohamed Al Marzouqi*[1]

[1] *Department of Chemical and Petroleum Engineering, United Arab Emirates University (UAEU), P.O.Box 15551, Al Ain, U.A.E*

[2] *Department of Chemical Engineering, Abu Dhabi University (ADU), P.O. Box 59911, Abu Dhabi, U.A.E.*

Abstract

The depletion of natural resources has dictated the need to seek alternative feedstock for the manufacture of high added-value products for industrial use. Here we present a road map for the manufacture of activated carbon submicron- and nanofibers from renewable and waste resources, namely lignin and recycled PET (r-PET). Using the electrospinning technique, mats of varying average fiber diameter were prepared, consisting of a varying mass ratio of lignin/r-PET. The carbonization of these mats proved that the morphology of the carbonized fibers depends on the average fiber diameter and on the lignin/r-PET mass ratio of the precursor ones. Using a 1/1 lignin/PET mass ratio results in the melting of the thinnest fibers (average diameter <300 nm). The melting of the thinnest fibers compromises their surface area, while the thicker ones retain their morphology and after activation their surface area can reach the value of 500 m^2/g. Using a higher lignin/PET mass ratio (at least 4/1) results in carbon nanofibers with average diameter lower than 200 nm, as the precursors retain their shape after carbonization.

Introduction

Carbon fibers are fibers with a carbon content of typically at least 90 % per unit mass [1]. They are an attractive form of carbon with remarkable properties that are suitable for the manufacture of composites, for energy storage devices and for adsorption and catalysis among other applications [1,2]. Their properties include the good thermal stability, high tensile strength and stiffness, and the ability to be functionalized [1,2]. When their diameter is in the nanoscale, then they are termed as carbon nanofibers (CNFs), and they are even more useful to applications where the large surface area is important. Their internal structure consists of amorphous and graphitic regions [1]. In general, their structure, morphology and properties depend on the method of manufacture, on their precursors and subsequent post-treatment.

One method of CNF production is the chemical vapor deposition, but it is rather cost prohibitive [3]. In contrast, the alternative method of thermal treatment of a suitable fibrous polymeric precursor seems to be more efficient and it is the main method of commercial carbon fiber production [1,4]. The most usual polymeric precursor for this purpose is poly(acrylonitrile) (PAN), while pitch is also commonly used [1,4]. Nevertheless, PAN is a rather expensive petroleum-based polymer, and its high cost limits the more widespread use of carbon fibers, since the cost of the precursor accounts for

* Address correspondence to Ioannis Zuburtikudis, E-mail: ioannis.zuburtikudis@adu.ac.ae

almost 50% of their manufacturing cost [4]. Therefore, there is a growing interest in the use of alternative, less expensive feedstock.

A promising alternative is lignin, the second most abundant natural polymer on earth behind cellulose [5]. This inexpensive, biorenewable feedstock comes as a by-product of the paper industry, while it can also be extracted from waste produced by agricultural activities. Its structure varies based on the plant source and the method of extraction, but typically consists of phenylpropane units (guaiacyl, syringyl and p-hydroxyphenyl propane) [6]. The growing interest in the utilization of lignin as a precursor for CNFs is justified by its relatively high carbon content (typically >60%), and its large availability [1]. For this purpose, it has been successfully used either alone, or blended with other polymers, such as poly(ethylene), poly(ethylene terephthalate), poly(vinyl alcohol), etc [1].

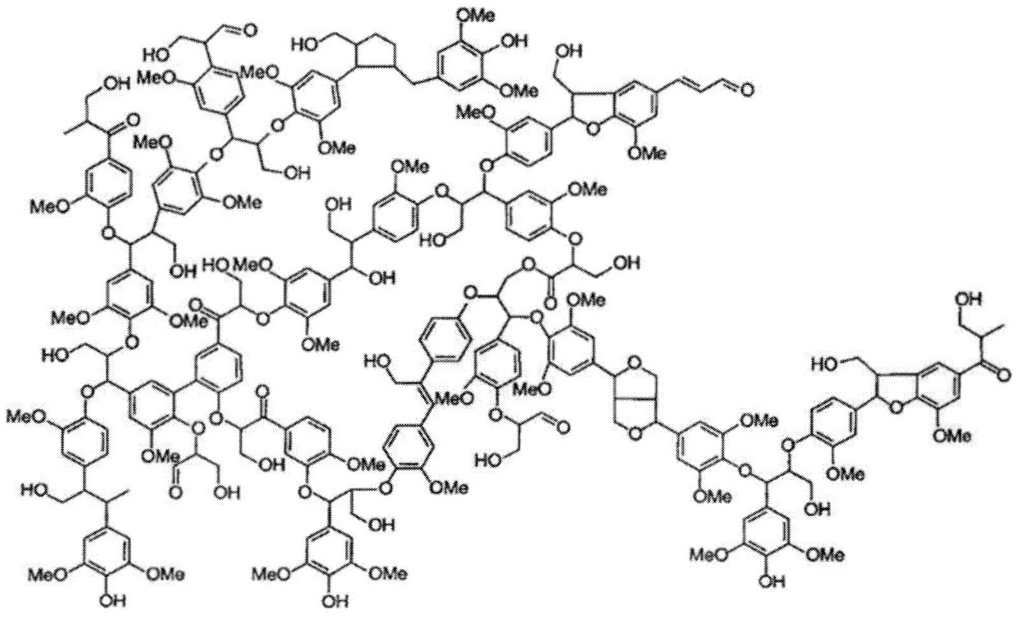

Figure 1. Representative structure of lignin [6]

Poly(ethylene terephthalate) (PET) on the other hand is one of the most common commodity polymers, widely used for the manufacture of many items of daily use, while it is one of the most recyclable plastic materials [7]. Thus, recycled PET is a low cost, readily available feedstock. Moreover, developing alternative uses for this raw material will have a positive environmental impact.

A common technique for producing submicron- and nano-fibers is electrospinning. The spinning of fibers with this simple technique is achieved when a polymer solution flows in an electrostatic field at a specific controlled rate through the tip of a syringe. The field helps in the stretching of the polymer jet, and after the evaporation of solvent, the fibers are formed and deposited on a collector. More information on this technique can be found in ref. 8. The most important feature of this method is that its process parameters can be easily adjusted; thus, a desirable fibrous morphology can be produced. Fibers from a wide variety of polymers and polymer blends have been successfully fabricated using this method.

For the preparation of CNFs, the polymeric precursor fibers fabricated through electrospinning undergo thermal treatment under inert atmosphere at high temperatures (typically >600 °C). During this carbonization process, thermal decomposition takes place, accompanied by the evolution of volatile gases (e.g. CO, CO_2, CH_4, CH_3CH_3, H_2, etc) [9]; the final product consists mainly of carbon, arranged in an amorphous polyaromatic structure with graphitic domains. In our work, lignin was combined with PET for the preparation of electrospun submicron- and nanofibers, which after carbonization yielded carbon submicron- and nanofibers.

Experimental part

The kraft lignin that we used here was supplied by Sigma-Aldritch (#471003, Mw~10,000 g/mol, low sulfonate content) and it was used without further treatment. The recycled PET (r-PET) was derived from waste water bottles, which were left to dry and were cut into small pieces. For the preparation of the electrospinning solutions, trifluoroacetic acid (99%) purchased from Merck was used. The solutions of lignin/r-PET were prepared and left overnight under magnetic stirring at room temperature until homogeneously dissolved.

The electrospinning process was carried out in a FUENCE E-sprayer (ES-2000S) apparatus in which the setup has vertical orientation. For the preparation of electrospun mats with a predetermined range of the average fiber diameter, we applied a design-of-experiments statistical methodology, in order to correlate the process and material parameters (spinning distance, voltage, flow rate, lignin/PET mass ratio, solution concentration) with the fibrous morphology. Details on the methodology that we followed for optimizing the process can be found elsewhere [8]. Briefly, we concluded that the most important factor affecting the average fiber diameter is the solution concentration.

For the purpose of examining the impact of the average fiber diameter as well as of the lignin/r-PET mass ratio on the morphology of the carbonized fibers, various electrospun samples were produced. Particularly, from lignin/r-PET with mass ratio 1/1, we produced various samples of different average fiber diameter, ranging from 80 to 781 nm. Moreover, we prepared electrospun fibers consisting of lignin/r-PET mass ratio of 1/1.5, having an average fiber diameter of 385 nm, and also a sample consisting of 4/1 mass ratio with an average fiber diameter of 92 nm.

The precursor electrospun fibers were carbonized under N_2 with a heating rate of 5 °C/min, until they reached 800 °C, where they were held for 1h. The process took place inside a temperature-programmed tubular furnace (Tube Furnace GSL-1500X-50, MTI).

The morphology of the electrospun and of the carbonized samples was investigated through scanning electron microscopy (SEM) (JEOL Neoscope JCM-5000). Thermogravimetric analysis was carried out under N_2 atmosphere, with a heating rate of 7 °C/min using a TA Q500 instrument.

Besides carbonization, activation of the carbonized samples took place at 600 °C for 1h, using CO_2 (flow rate ~500 – 600 cc/min). Their BET surface area was measured through N_2 adsorption using a Micromeritics - TriStar II Plus analyser.

Results and discussion

Our efforts to fabricate electrospun fibers from lignin alone were fruitless, as lignin has a relatively low molecular weight; in this case only electrospray was produced. Therefore, the presence of PET is necessary for fiber formation. The aim of our research is to use the CNFs for adsorption, so, the fabrication of nanofibers with the minimum possible diameter is necessary, as in this case the surface area is expected to be maximized. Here, using lignin/r-PET with mass ratio 1/1 gave nanofibers with average fiber diameter of 80 nm; these were the thinnest nanofibers produced successfully.

Repeated attempts, however, to carbonize fibers of different average diameter indicated that the morphology of CNFs depends on the average fiber diameter of the precursor fibers. Precursor fibers of lignin/r-PET ratio of 1/1 with relatively large average diameter (>300 nm) yield carbon submicron fibers with similar diameter as the precursor ones, retaining their morphology. In contrast, as the precursor average diameter is lowered in the nanoscale, and especially for fibers of the minimum diameter (<100 nm), the fibrous morphology is compromised as the fibers exhibit extensive melting during carbonization, and they merge with each other. In figure 2, two representative cases are presented, which illuminate this phenomenon; the carbonization outcome for fibers with average diameter of 121 nm, and for those of 547 nm.

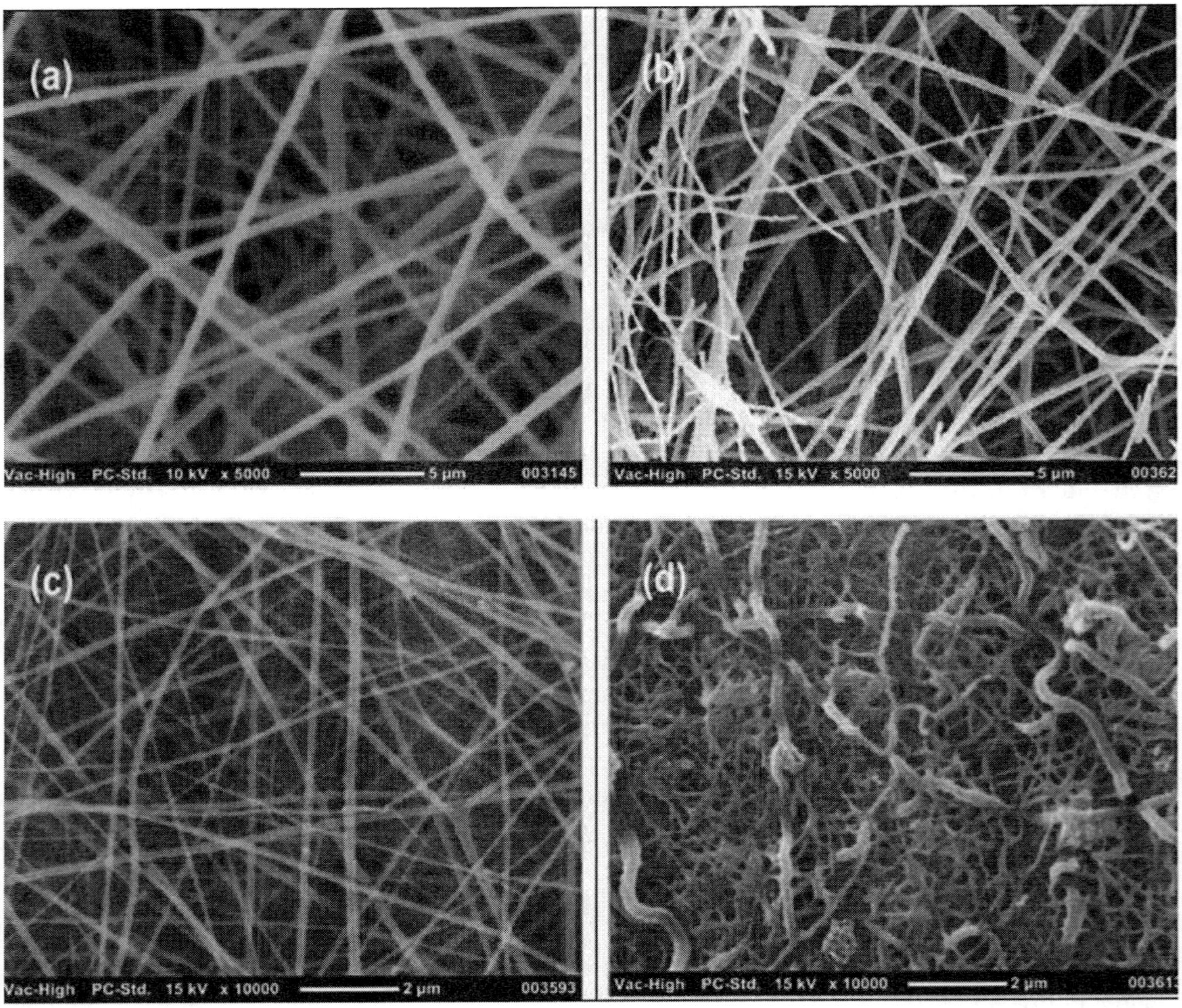

Figure 2. Electrospun fibers with average diameter 547 nm (a) and their corresponding carbonized ones (b); electrospun fibers with average diameter 121 nm (c) and their corresponding carbonized ones (d)

While PET is semicrystalline showing a melting peak at around 251 °C, lignin is amorphous [10]; when these two polymers are mixed, the blend can be either amorphous or semicrystalline, depending on the relative mass percentage of lignin [10]. When the percentage of lignin is less than 35 % w/w, the fibers formed from the blend exhibit a melting peak in the region of 242 – 250 °C. In contrast, when the ratio of lignin is higher than 50 %, the lignin chains disrupt the PET crystallites to such an extent that the blend doesn't exhibit a detectable melting peak [10]. However, it seems that this depends also on the fiber diameter, and the decomposition of lignin plays a role here. Lignin decomposes over a wide temperature range, in various stages [11]. The first stage of lignin decomposition appears in the region 200 – 250 °C, with a peak at 237 °C (figure 3). Results of the thermogravimetric analysis, showed that the weight loss of the electrospun mats in this region depends on the average fiber diameter. Mats consisting of relatively thick fibers (average diameter 300 - 800 nm) show relatively low and almost stable weight loss in this region (~ 8.5 – 9 %/°C). Nevertheless, the fibers consisting of thinner fibers show a higher decomposition rate and weight loss, which reaches the value of 11.3 %/°C for the electrospun mat of 121 nm, and the value of 14.5 %/°C for the fibers of the minimum average diameter (80 nm). The reason is that the thinnest fibers have many more lignin macromolecules located at or near the fiber surface, thus the heat and mass transfer doesn't restrict their decomposition to a large extent. This translates into more lignin macromolecules decomposing, leaving more space for PET chains to move, therefore, there is apparent melting of the thinnest fibers in the region of PET melting (~250 °C).

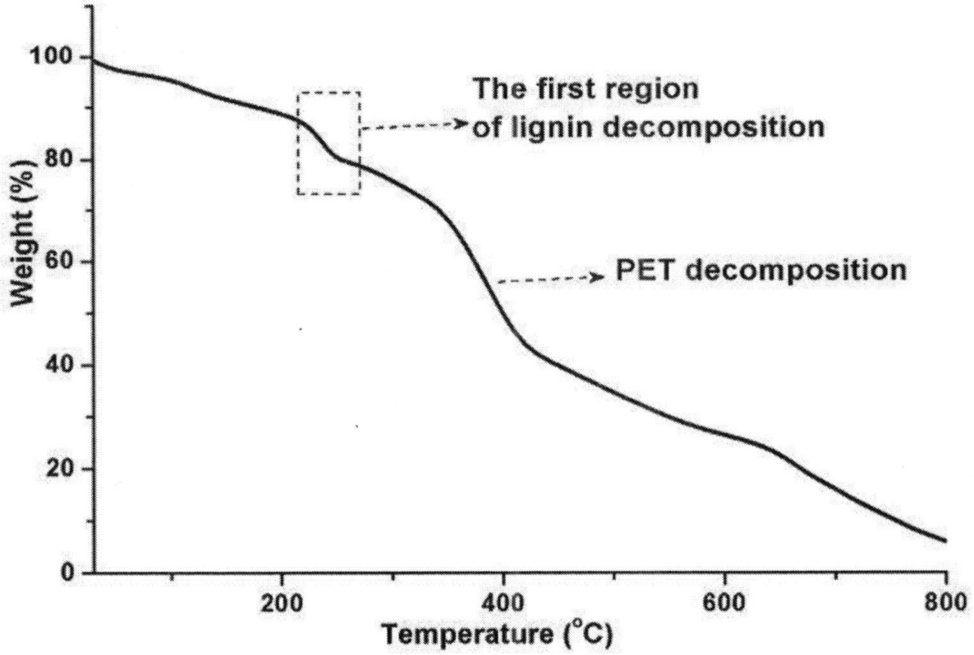

Figure 3. Representative TGA curve for lignin/r-PET fibers of 1/1 mas ratio

Moreover, the outcome of the carbonization seems to depend on the lignin/r-PET mass ratio, as well. For fibers consisting of lignin/r-PET mass ratio of 1/1.5, it seems that the melting occurs even at relatively thicker fibers (average diameter of 385 nm) (figure 4(a) – 4(b)). In contrast, increasing the ratio to lignin/r-PET=4/1, the melting is much less pronounced even for electrospun mats consisting of fibers with average diameter of 167 nm (figure 4(c) – (d)). The effect of the lignin/r-PET mass ratio in the carbonization outcome is still under investigation.

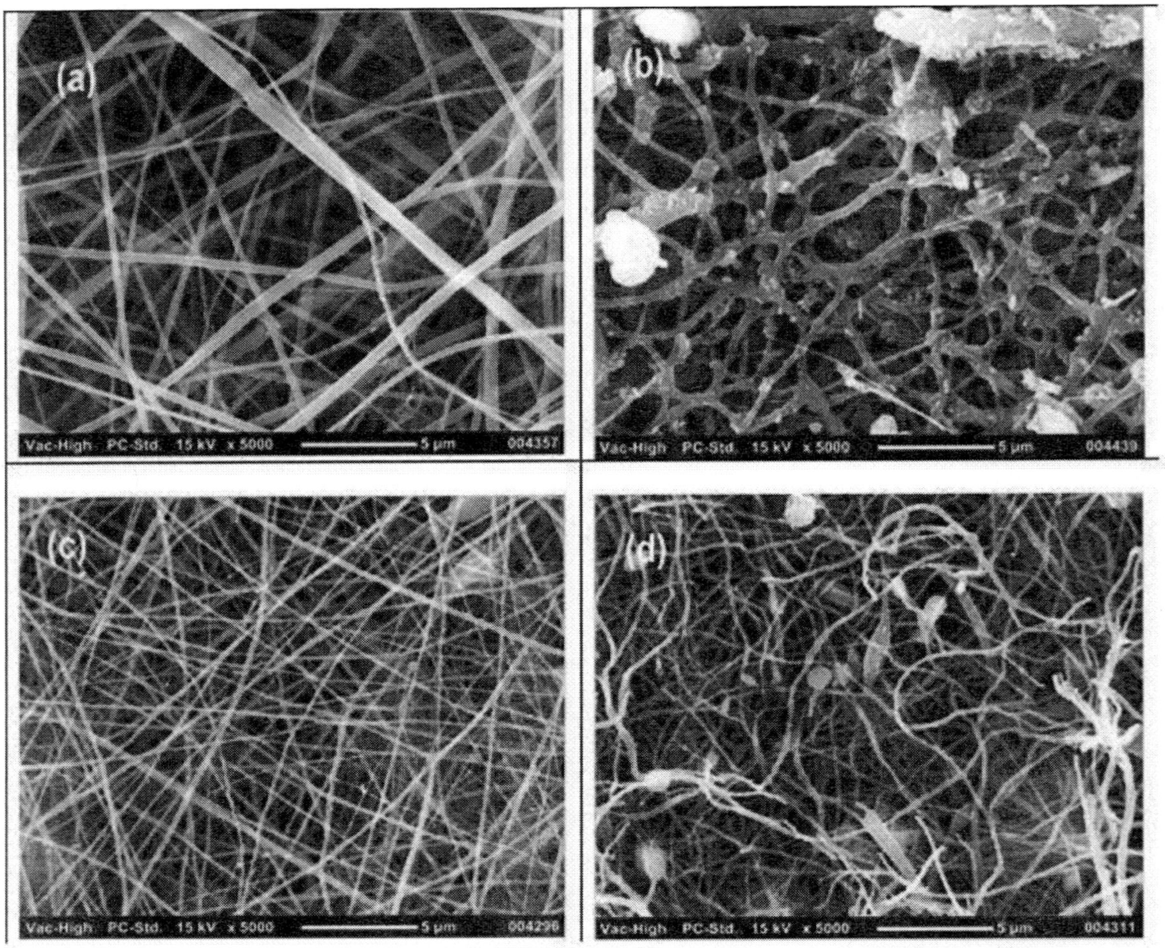

Figure 4. Electrospun fibers of lignin/r-PET=1/1.5 mass ratio (a) and their corresponding carbonized ones (b); electrospun fibers of lignin/r-PET=4/1 mass ratio (c) and their corresponding carbonized ones (d)

As it was mentioned previously, we intend to use the activated CNFs for adsorption, therefore, it is important to maximize their surface area and porosity. Our preliminary results for the activated CNFs suggest that our assumption that the minimization of the average diameter of the precursor fibers maximizes the surface area of the carbonized ones is practically refuted by the fact that the thinnest fibers merge with each other during carbonization. Reducing the average diameter up to a certain point indeed increases the surface area, but further reduction compromises the surface area due to melting. This behavior for the fibers consisting of 1/1 lignin/r-PET is presented in figure 5, in which the maximum surface area for the corresponding activated CNFs can reach the value of 500 m^2/g. This phenomenon is still under examination.

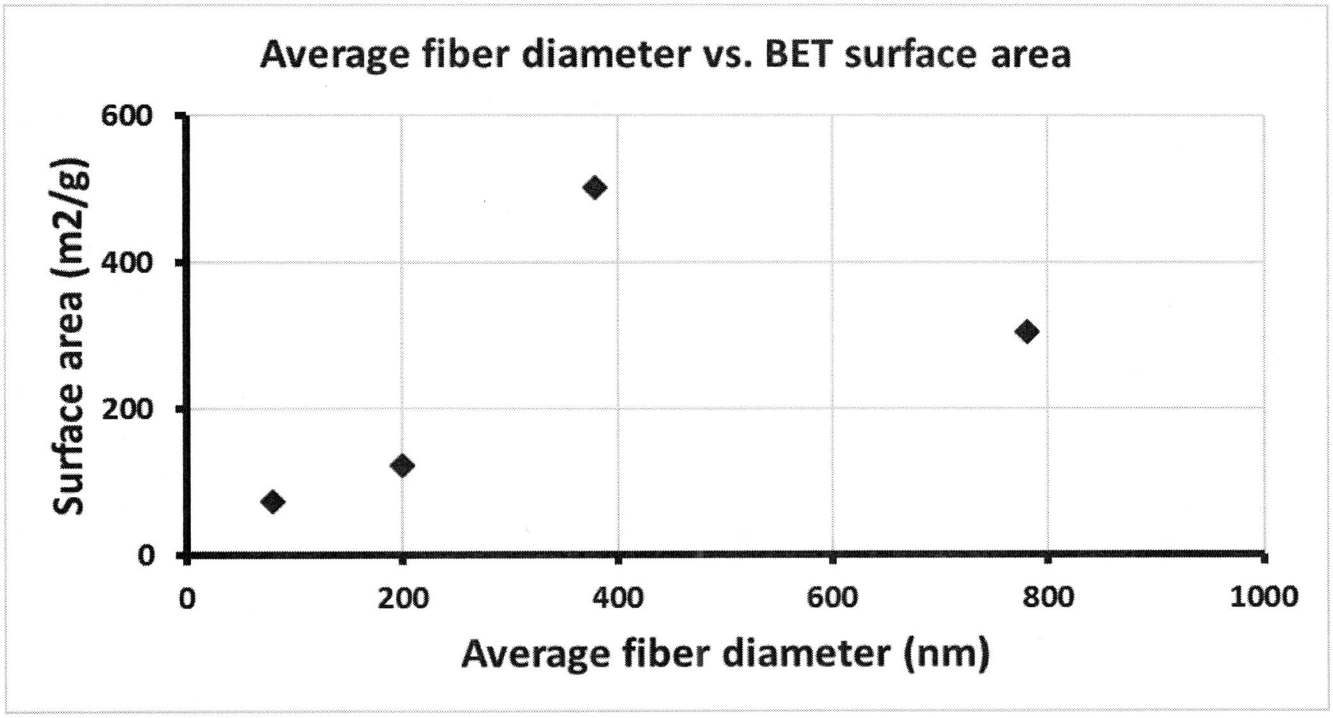

Figure 5. Average fiber diameter vs. BET surface area for the activated CNFs produced from precursor fibers of lignin/PET with mass ratio 1/1

Conclusions

In this work, we have demonstrated the feasibility of producing activated carbon fibers with significant values of specific surface area using renewable and inexpensive resources, namely lignin and recycled PET. Precursor fibers were fabricated using the electrospinning technique, in which the desirable fiber diameter and morphology can be adjusted. Carbonization of the precursor electrospun fibers shows that the carbonization outcome depends on the average fiber diameter and on the mass ratio of the precursor ones. Using a high lignin/r-PET mass ratio renders feasible the preparation of activated carbon nanofibers with average diameter of less than 200 nm. The results presented here illuminate the factors affecting the successful fabrication of carbon fibers from waste resources in the framework of sustainability and they highlight the aspects that should be taken into consideration in processes involving the transfer of heat and mass, such as carbonization. Therefore, they present a route for the manufacture of tailor-made adsorbents with significant surface area from low-cost feedstock.

References

1. E. Frank, L. M. Steudle, D. Ingildeev, J. M. Spörl, and M. R. Buchmeiser, Carbon Fibers: Precursor Systems, Processing, Structure, and Properties. Angew. Chemie Int. Ed. 53, 5262 (2014)
2. A. Ogale, J. Zhang Meng, and J. Jin, Recent advances in carbon fibers derived from biobased precursors. J. Appl. Polym. Sci. 133, 713 (2016)

3. Gu SY, Ren J, Vancso GJ. Process optimization and empirical modeling for electrospun polyacrylonitrile (PAN) nanofiber precursor of carbon nanofibers. European Polymer Journal 41, 2559–2568 (2005)
4. Baker DA, Rials TG. Recent Advances in Low-Cost Carbon Fiber Manufacture from Lignin. J. Appl. Polym. Sci. 130 (2), 713-728 (2013)
5. Duval A, Lawoko M. A review on lignin-based polymeric, micro- and nano-structured materials. Reactive & Functional Polymers 85, 78–96 (2014)
6. Wool RP. Lignin polymers and composites. In: Wool RP, Sun XS, Biobased polymers and composites, 1st ed., Elsevier 2005, p. 551-598
7. Wang CQ, Wang H, Liu YN. Separation of polyethylene terephthalate from municipal waste plastics by froth flotation for recycling industry. Waste Management 35, 42–47, (2015)
8. E. Svinterikos and I. Zuburtikudis, Tailor-made electrospun nanofibers of biowaste lignin/recycled poly(ethylene terephthalate). J. Polym. Environ. 25, 465 (2017).
9. C. Rodríguez Correa, M. Stollovsky, T. Hehr, Y. Rauscher, B. Rolli, and A. Kruse, Influence of the Carbonization Process on Activated Carbon Properties from Lignin and Lignin-Rich Biomasses. ACS Sustain. Chem. Eng. 5, 8222 (2017).
10. Svinterikos E and Zuburtikudis I, Carbon nanofibers from renewable bioresources (lignin) and a recycled commodity polymer [poly(ethylene terephthalate)], J. Appl. Polym. Sci. 133, 43936, (2016)
11. C. Rodríguez Correa, M. Stollovsky, T. Hehr, Y. Rauscher, B. Rolli, and A. Kruse, Influence of the Carbonization Process on Activated Carbon Properties from Lignin and Lignin-Rich Biomasses. ACS Sustain. Chem. Eng. 5, 8222 (2017)

539328 Supercritical Fluids As Reaction Media for Scalable Production of Carbon Nanomaterials

Abstract:

We have demonstrated scalable synthesis of a variety of nanocarbons including carbon nanotubes (CNTs), carbon nanofibers (CNFs), and onion-like carbon (OLC) in a batch reactor using supercritical fluids as a reaction media. The process utilizes toluene, ethanol, or butanol as a carbon precursor in combination with ferrocene that serves the dual role of a catalyst precursor and a secondary carbon source. Growth with supercritical toluene at 600ºC in the absence of water yield large-diameter CNTs and CNFs while the introduction of 0.05 ml of water enhances the product yield by 50% and promotes the formation of smaller-diameter CNTs. The exterior surface of CNTs was decorated with highly monodispersed Fe nanoparticles. At lower temperature (400ºC) in the absence of water, supercritical toluene produces mainly OLC; similarly, high OLC selectivity was achieved by increasing the amount of added water to 5.55 mol/L. The tendency for alcohols to produce tubular structures increases in the order $C_2H_5OH < C_3H_7OH < C_4H_9OH$, which suggests a direct correlation between the aliphatic chain length of alcohol and the tendency to form a tubular structure. The use of supercritical fluids for growth does not only provide a route for selective and scalable growth of a variety of carbon nanomaterials, but also provides a unique one-step approach that is free of aggressive acid treatment for synthesis of CNT-supported or encapsulated Fe composites with potential applications in catalysis and energy storage.

537450 From Carbon Nanotube Liquid Crystalline Solutions to Functional Fibers

Abstract:

Carbon nanotubes (CNTs) have a remarkable combination of electrical, thermal, and mechanical properties. In order to realize these properties in applications, we need to understand and control the behavior of fluid phases of these nanomaterials. Fluid phase assembly is among the most industrially attractive routes for scalable CNT processing into macromaterials such as multifunctional fibers. Chlorosulfonic acid (CSA) is known as an ideal solvent for CNTs, which spontaneously dissolves them without compromising properties. High concentration solution of CNTs in CSA forms a nematic liquid crystalline phase characterized by a long-range orientational order. The natural orientational alignment of CNTs within the nematic phase would highly benefit the alignment of CNTs in fluid phase-processed macroscopic fibers, improving the electrical and mechanical properties in the alignment direction. Wet-spun CNT fibers combine the specific strength, stiffness, and thermal conductivity of carbon fibers with the specific electrical conductivity of metals. Given these properties, CNT fibers can serve as an attractive structural support on to which functional materials can be deposited to develop active optoelectronics. Using light-emission as an example, in this talk we will discuss our approach to transform carbon nanotube fibers to active devices.

535977 Metal Organic Frameworks Promise High Activity and Stability of Carbonic Anhydrase in Synthetic Environment

Abstract:

Carbonic anhydrase (CA) has shown great potential for carbon dioxide (CO_2) sequestration and transformation due to its ultrahigh turnover. However, the fragile nature of the free enzyme significantly limits its application in industry. Herein we propose to develop robust CA-based biocatalyst for CO_2 sequestration and transformation based on metal organic frameworks (MOFs). For this, we employed a hydrophilic and a hydrophobic MOF support for CA immobilization and evaluated the support characteristics that will ensure maximum functionality and stability of the enzyme. MOFs' morphology, crystallinity and particle size were evaluated by scanning electron microscopy (SEM), dynamic light scattering (DLS) and X-Ray diffraction (XRD) respectively and compared with the characteristics of the conjugates. Analysis showed that the supports influenced both the enzyme loading and activity, as well as determined enzyme kinetics and stability at interfaces. Further, results showed that controlling the hydrophilicity of the support could lead to not only increased enzyme functionality but further, increased stability relative to that of free enzyme. These results point to MOFs having a high potential to fill in the gap for developing sensitive supports with tunable hydration conditions that allow for high enzyme activity and stability retention and hint at extended implementation capability when CO_2 storage and transformation based on enzyme-imposed green technologies is considered.

538346 Estrogen Receptor-Targeted Multiplexing Photoacoustic Polymeric Nanoparticles for Diagnostic and Treatment of Breast Cancer

Abstract:

Estrogen Receptor-Targeted Multiplexing Photoacoustic Polymeric Nanoparticles for Diagnostic and Treatment of Breast Cancer

C.Nino-Vargas[1], E.Blatchford-Rodriguez[1], Z.Begnell[1], R.Khalid[1], G.Petruncio[2], M.Paige[2], C.Salvador-Morales[1]

[1]Bioengineering Department, George Mason University, 4400 University Drive, Fairfax, Virginia, 22030,
[2]Chemistry Department, George Mason University, 4400 University Drive, Fairfax, Virginia, 22030.

Introduction. Breast cancer is the second cause of death in women after lung cancer, with approximately 41,000 breast cancer deaths expected this year [1]. Systemic chemotherapy is one of the most common approaches to therapy, but its systemic toxicity is a major drawback that limits chemotherapeutic utility and effectiveness [2]. Another significant problem is the generation of the multidrug resistance phenomenon, which, among other factors, arises from inhomogeneous distribution of the drug within the tumor [3]. A recent focus of cancer research is on developments of theranostic systems capable of providing diagnostic imaging and therapeutic functions simultaneously. An emerging scientific and clinical consensus is that theranostic devices may provide several medical benefits including: a) molecular imaging functionalities to help physicians track drug distribution within the solid tumor, b) valuable guidance on selection of a safe and efficacious dose, and c) real-time objective monitoring of therapeutic response. The multiplexing photoacoustic capability of these nanovehicles will render substantial anatomical, physiological and functional information of the breast cancer tumor.

Here, we report the design and synthesis of a novel targeted cancer theranostics device for breast cancer treatment that leverages the highly tunable internal properties of Patchy Polymeric Nanoparticles (PPNPs). Estrogen Receptor-Targeted Patchy Polymeric Nanoparticles loaded with Indocyanine Green (ICG) and IR-26 consist of poly (lactic-co-glycolic acid) (PLGA) and lipid-PEGylated-Estrone. PLGA and ICG are biocompatible and FDA-approved biomaterials. These nanoparticles have unique patch-core-shell structural features: hollow core and patchy surface. The particle's patchy surface can be utilized to enhance tumor targeting because of the high concentration of targeting moieties in one region of the particle's surface.

Materials and Methods. The surface of our nanoparticles is functionalized with Lipid-PEGylated-Estrone molecules that will bind specifically to estrogen receptors that are over-expressed in breast cancer cells. We performed a comprehensive characterization of the targeting moiety using Nuclear Magnetic Resonance (NMR), Matrix Assisted Laser Desorption/Ionization (MALDI) and High Performance Liquid Chromatography (HPLC). Our results show that we successfully synthesized the targeting moiety. Next, we synthesized the targeted nanoparticles using a modified single emulsion method previously published [4]. Moreover, we characterized these nanoparticles with different experimental techniques including Dynamic Light Scattering (DLS), Transmission Electron Microscopy (TEM), Absorption and Near-IR Spectroscopy. Our results show that these particles have an average hydrodynamic diameter of 165 nm ± 5 nm. Also, the particles' size is stable for months in frozen and lyophilized storage conditions. Furthermore, we assessed the absorption ability of these nanoparticles in the Near-IR region of the electromagnetic spectrum. These nanoparticles present an absorption peak at 786 nm and 895 nm wavelength. TEM shows the spherical morphology of these nanoparticles. Additionally, we evaluated the cellular toxicity of the engineered nanoparticles at different concentrations. We observed that these nanoparticles are not toxic as expected since PLGA and ICG are biocompatible, and IR is enveloped by the polymeric matrix. Moreover, we evaluated the cell uptake of untargeted and targeted nanoparticles at 2 hrs. and 24 hrs. time points using T-47D breast cancer cells. We observed that targeted nanoparticles are avidly taken by breast cancer cells at both 2hrs and 24 hrs. However, there cell uptake at 24 hrs. is higher than that at 2hrs. The assessment of the photoacoustic performance and *in vitro* cellular uptake of these nanoparticles is underway. Future experiments will be carried out to evaluate their in *vivo* photoacoustic performance. Equally important, we have carried out drug release studies to determine the drug release profile of these nanoparticles. Our most recent studies demonstrate that these nanoparticles release 70% of the drug over a month.

Conclusions: These preliminary results show the feasibility of engineering multiplexing photoacoustic theranostic carriers for breast cancer treatment. Thus, these theranostics have the potential to provide meaningful information about the tumor vasculature because of their ability to penetrate deep into tissue while killing at the same time tumor cells selectively.

537430 Engineering of Charge Transfer Complex Nanocrystals By Electrocrystallization

Abstract:

Crystal structure and morphology of nanoparticles determine their interfacial atomic arrangements and thus control their properties. To date, controlling the morphology of the nanoparticles is still a major barrier that greatly limits their applications. This work discusses the experimental and theoretical methods to study and control the morphology of the nanocrystals. Guided by the thermodynamics and kinetic transition theory, we used electrocrystallization to make K(def)TCP nanocrystals and tuned the electrochemical parameters to alter the nanocrystal morphologies. Our results suggest that the electrochemical parameters influence the K(def)TCP nanocrystal morphology. We obtained nanoparticle shapes ranging from nanorods to rhombohedral shape depending on the solution concentration, the applied potential and the current. We modeled the observed crystal growth behavior with a Monte-Carlo based computer simulation. The simulation results qualitatively match the experimental findings. This work contributes to the understanding of the crystal growth behavior and the thermodynamic/kinetic morphology transition in electrocrystallization. We will show the application of the engineered K(def)TCP nanocrystals in gas sensing.

539030 Silica-Coated, Near-UV Activated $YVO_4:Eu^{3+},Bi^{3+}$ Nanophosphors for Dynamic Cell Imaging

Abstract:

Luminescent rare-earth-based inorganic nanoparticles (nanophosphors) are promising bioimaging agents due to their high photostability, sharp emission bands and relatively low toxicity [1]. Flame aerosol technology provides a scalable and highly reproducible process for production of such nanophosphors (*e.g.* $Y_2O_3:Eu^{3+},Tb^{3+}$) with precise control of their composition and properties [2,3]. Nanophosphors that can be excited in the near-ultraviolet and visible region, such as $YVO_4:Eu^{3+},Bi^{3+}$, provide a useful tool for bioimaging and *in vitro* dosimetry studies using conventional fluorescence microscopes.

Here, $YVO_4:Eu^{3+},Bi^{3+}$ nanophosphors are made by flame spray pyrolysis and in-situ coated by a nanothin amorphous silica coating. The optimal Bi content for maximum red-shift of their excitation band edge towards the visible region is identified through systematic experiments. The nanophosphors with the optimal composition are highly crystalline and appear bright red under a conventional fluorescence microscope. Their photostability during dynamic imaging of HeLa cells *in vitro* is confirmed, contrary to commercial fluorescent (organic-dye labeled) SiO_2 nanoparticles that exhibit 50% photobleaching within 3.5 h [4]

The potential of these nanophosphors for dynamic cell imaging is demonstrated and compared to conventional organic dyes, demonstrating the nanophosphor superior performance/ We employed cell-bacterial co-cultures to identify how GFP-expressing *S. pneumoniae* bacteria interact with human epithelial lung cancer cells (A549). Furthermore, the nanoparticle interaction with these cells is further examined by employing upright and inverted cell culture orientations, and by the surface biofunctionalization of the nanoparticles with targeting antibodies compared with non-targeting bovine serum albumin. The amorphous silica coating here facilitates the biofunctionalization of the nanoparticle surface. The developed silica coated nanoprobes here will open up several avenues on dynamic nanoparticle-cell imaging.

REFERENCES

1. A. Escudero, C. Carrillo-Carrion, M. V. Zyuzin & W. J. Parak. Luminescent rare-earth-based nanoparticles: A summarized overview of their synthesis, functionalization, and applications. *Topics Curr. Chem.* **374**, (2016).
2. G. A. Sotiriou, M. Schneider & S. E. Pratsinis. Color-tunable nanophosphors by codoping flame-made Y_2O_3 with Tb and Eu. *J. Phys. Chem. C* **115**, 1084-1089 (2011).
3. G. A. Sotiriou, D. Franco, D. Poulikakos & A. Ferrari. Optically stable biocompatible flame-made SiO_2-coated Y_2O_3: Tb^{3+} nanophosphors for cell imaging. *ACS Nano* **6**, 3888-3897 (2012).
4. A. Spyrogianni, P. Tiefenboeck, F. H. L. Starsich, K. Keevend, F. Krumeich, I. K. Herrmann, J. C. Leroux & G. A. Sotiriou. "Near UV Activated, Photostable Nanophosphors for *in vitro*Dosimetry and Dynamic Bioimaging" *AIChE J.***in press**, DOI: 10.1002/aic.16166 (2018).

539060 Stimuli Responsive Nano-Agents: From Drug Delivery to Oil and Gas Industry

Abstract:

Several concepts that have been successfully used for targeted drug delivery for medical diagnostics and treatment can lend themselves to some key oil and gas applications, despite the clear differences in the systems and environments involved. An example of such concepts is the utilization of stimuli responsive materials to deliver treatment or image contrast agents to specific targets in the human body, or the oil reservoir, without the side effects and unnecessary costs associated with excess release or delivery of agents to where they are not needed.

The Responsive Reservoir Nano-agents Program was initiated at Saudi Aramcoâ€™s upstream Advanced Research Center in 2016. The main objective of this program is to develop nano-agents that can travel deep into the reservoir to collect reservoir information or deliver chemical treatments to *hard-to-reach* regions, autonomously by exploiting the induced or natural chemical and physical conditions in the reservoir. These regions are hardly accessed by conventional water or gas flooding techniques.

In this talk, a proof-of-concept on the effect of salt and surfactant gradients on the migration of reservoir nano-agents will be presented. The experiments use novel microfluidic devices and direct microscopic visualization to quantify the migration of solid colloidal particles ($<1\,\mu m$) under different gradients and to understand the underpinning electrokinetic and interfacial phenomena. The results demonstrate the ability of salinity and surfactant gradients to transport reservoir nano-agents up or down the imposed gradients in a predictable way. In this way, natural or induced gradients in the reservoir can be exploited to direct the reservoir nano-agents to certain regions in the reservoir for discovering or recovering the trapped resources.

526017 Directed Assembly of Polarizable Nanoparticles

Abstract:

Directed assembly of dielectric and paramagnetic nanoparticles can be used to synthesize diverse functional materials that polarize in response to an externally applied electric or magnetic field. However, theories capable of predicting the self-assembled states are lacking. In the proposed work, we develop a complete thermodynamic description of such assemblies built from spherical nanoparticles. We show how an important physical feature of these types of particles, mutual polarization, sculpts the free energy landscape and has a remarkably strong influence on the nature of the self-assembled states. Modeling the mutual polarization among nanoparticles requires solving a many-bodied problem for the particle dipole moments. Typically, this computationally expensive task is avoided by neglecting mutual polarization and assuming each particle in a concentrated dispersion acquires the same dipole moment as a single, isolated particle. This constant dipole assumption leads to qualitatively incorrect predictions that fail to satisfy specific thermodynamic criteria for coexisting phases in equilibrium. Correctly accounting for mutual polarization enables a thermodynamic theory that describes the equilibrium phase diagram of polarizable dispersions in terms of experimentally controllable variables. Our theoretical predictions agree with the phase behavior we observe in dynamic simulations of these dispersions as well as that in experiments of field-directed structural transitions. In contrast to predictions of a constant dipole model, we find that dispersions of particles with different dielectric constant or magnetic permeability exhibit qualitatively different phase behavior. This new model also predicts the existence of a eutectic point at which two crystalline phases and a disordered phase of nanoparticles all simultaneously coexist.

532799 *Electrochemical Etching and Oxidation Stability of Mxene Nanosheets*

Abstract:

MXenes are a new group of 2D nanosheets that hold promise for a range of functional material applications. They are derived from MAX phases, which have the chemical composition $M_{n+1}AX_n$, where n = 1, 2 or 3 (M_2AX, M_3X_2, or M_4AX_3), “M” is an early transition metal, “A” is an A group (mostly groups 13 and 14) element, and “X” is C and/or N. Highly hazardous HF or LiF/HCl solution is widely used to remove A layer from parent MAX phase to get MXenes with a combination of -F, -OH, -O, or/and -Cl terminal groups. Here we successfully demonstrate the electrochemical etching of Al from porous Ti_2AlC in dilute HCl solution, with no use of toxic fluoride solutions. This approach is a less hazardous method to process and handle MXenes compared to prior works. We also observed that electrochemical etching can result in subsequent over-etching to carbide-derived carbon. In addition, we studied MXene oxidation in various media. We observed 1) faster oxidation in liquid media than in solid media, 2) continuous oxidation in pure MXene films and MXene/polymer composites, and 3) accelerated MXene film oxidation under UV irradiation.

538776 Antigen-Antibody Nanoparticle Bioconjugates and Their Polymorphs

Abstract:

In recent years, the integration of nanomaterials with proteins or DNA has been the focus of intense research. Such materials have the advantage of combining the unique properties of nanomaterials with the highly specific binding mechanisms of biological systems. This has opened the door for many novel and exciting applications as biosensors, and to the design of new hybrid materials through their assembly in 2D and 3D ordered structures. We focus here on a specific type of nanoparticle bioconjugates, that relies on antigen-antibody interactions to form 3D hybrid crystalline structures. For this purpose, we develop a coarse-grained model for nanoparticles grafted with antibody molecules and their binding with antigens. In particular, we isolate two possible states for antigen-antibody pairs during the binding process, termed as recognition and anchoring states. Using molecular simulation, we calculate the thermodynamic and structural features of two possible crystal structures or polymorphs, the body-centered cubic and simple cubic phases, and of the melt. This leads us to determine the phase diagram and domain of stability of the two solid phases. In particular, the role played by the switching process between anchoring and recognition states during melting is identified, shedding light on the complex microscopic mechanisms in these systems.

531071 Adsorption and Denaturation of Polymeric Nanoparticles at an Interface

Abstract:

Adsorption and denaturation of polymeric nanoparticles at an interface

Chang Tian[1], Jie Feng[1], Robert K. Prudâ€™homme[1]

[1] Department of Chemical and Biological Engineering, Princeton University, Princeton, New Jersey, 08540

Understanding nanoparticle (NP) self-assembly at an interface is essential to control NP adsorption onto targeted areas and determine subsequent drug release kinetics. However, few studies have characterized the adsorption of NPs onto interfaces quantitatively. By looking at the dynamic surface tension change over time, we investigate air-water interfacial adsorption of NPs, which are made with a hydrophobic core and hydrophilic polyethylene glycol corona through the Flash NanoPrecipitation (FNP) process. By utilizing the maximum bubble pressure tensiometer, we are able to record surface tension over the scales from 100 ms to 30 s. Three stages were documented for NP adsorption, which includes an early stage of free diffusion, a later one with steric adsorption barriers and a hitherto unrealized region where adsorption energy sharply increases due to surface restructuring of NPs to expose their hydrophobic cores. By modifying the Ward and Tordai model, we collapsed curves of surface tension change for NPs of different concentrations and physical characteristics into one dimensionless â€œmaster curveâ€ via time-concentration shifting. In addition, we have analyzed the kinetics of NP adsorption in the presence of types of surfactants. Surfactants usually decrease the surface energy of an interface. However, we have observed phenomena where surfactants slow down the NP adsorption kinetics so as to raise the surface tension compared to that of a system with NPs alone. We have shown that this change of adsorption kinetics depends on the surfactant charge (anionic, cationic or non-ionic) as well as the formulation of NPs. A detailed mechanism is proposed to explain this NP-surfactant interaction.

533105 Driving Forces for Oriented Aggregation-Based Crystallization and Assembly

Abstract:

Driving forces for oriented aggregation-based crystallization and assembly depend on azimuthal crystallographic alignment between particles. The types of driving forces that can be sensitive to relative orientation include Coulombic, van der Waals (vdW), solvation, and ion correlation forces. The theoretical underpinnings of these anisotropic forces are well established, but techniques that can directly measure them for a given pair of interacting oriented crystal faces have generally been limited to use of macroscopic yet atomically flat single crystals. We report measurement of anisotropic forces between rutile TiO_2 (001) nanocrystals as a function of their azimuthal orientation and surface hydration extent using a combined environmental transmission electron microscopy-atomic force microscopy (AFM) technique. Atomically flat rutile TiO_2 (001) AFM tips and opposing rutile TiO_2 (001) substrates were fabricated by focused ion beam milling to excise nanocrystals from the surface of a single monolith that was pre-oriented, cut, and polished to prepare the (001) face. At tens of nanometers of separation, the attractive forces are weak and show no dependence on azimuthal alignment nor surface hydration. At separations of approximately one hydration layer, attractive forces are strongly dependent on azimuthal alignment and systematically decrease as intervening water density increases. Measured forces closely agree with predictions from Lifshitz theory and show that dispersion forces are capable of generating a torque between particles interacting in solution and between grains in materials.

533760 Rheo-Electric Behavior of Carbon Black Suspensions in Shear Flow

Abstract:

Carbon black is a ubiquitous additive used to modify the optical, mechanical and electrical properties of polymers, fluids and gels. There is recent renewed interest in the properties of specialty, high-structured, conductive carbon blacks due to their use as the electrical current collector in electrochemical flow technologies. Despite its important role in system performance, there is little understanding of the electron transfer dynamics in such materials particularly when subjected to transient and steady shear flows.

In this work, the rheo-electric behavior of a high-structured, conductive carbon black (Vulcan XC-72) is studied in mineral oil. Mineral oil is a model dielectric fluid that exhibits two important characteristics for this study: 1.) its viscosity is sufficient to induce significant structural erosion at shear rates accessible in commercial rheometers, and 2.) it is not ionogenic so that the electrical transport properties of the carbon black can be isolated from double-layer capacitance. The rheological and electrical response of the suspension under steady shear flow is thoroughly characterized. These measurements are complemented with small angle neutron scattering measurements made in a commercial stress-controlled rheometer (RheoSANS). We find that these suspensions, when sheared above their macroscopic yield stress, undergo shear-thinning and reversible agglomerate erosion. This structural erosion has a dramatic effect on the electrical properties of the suspension and the resulting complex impedance is fit to an equivalent circuit that describes the evolution of the charge transfer dynamics. This model reveals the important role of colloidal forces in the electron transport process in such suspensions.

541565 25 by 25: Chemical Engineering in the Next 25 Years

Abstract:

In 2008, the centennial year for AIChE, we asked 25 chemical engineers for their visions of the profession from around the US and the world: Nine faculty, eight from industry, and eight post-docs and graduate students. We posed four questions related to each individual's industrial sector or area of research, asking for extrapolation, new impacts on existing sectors, new sectors, and an optional, more open-ended comment on the future of the profession as a whole.

In this 110th year of AIChE, we have revisited these questions with many of the same people. Ten years has seen many developments in our profession, in society, and in our thought-leaders' experiences. Here we report on their perspectives, comparing them to the 2008 visions.

541542 The Future of Chemical Engineering Itself

Abstract:

Chemical engineeringâ€™s future is bright because it is vital to so much of society and industry -- and its future is complicated by its rapidly expanding universe of activities. It must represent depth in core skills centered on the chemical sciences including biology and materials. At the same time, it must be transdisciplinary; it must exemplify the need and value of reaching across disciplines in borderless ways.

One way is to create and disseminate a compelling vision and strategy for chemical engineering, research, and education for the future. AIChE and the National Academies are working together to that end. In 1988, the National Academies published "Frontiers in Chemical Engineering: Research Needs and Opportunities.â€ Its committee of authors outlined a wide-ranging roadmap to turn promising research opportunities into reality, while calling for researchers to embrace new frontiers.

Since 1988, that has happened -- along with many other transformative changes like the Web, fracking, personalized medicine, and social media. A new nationwide effort, aided by international contributions, will address both the identity of chemical engineering and its future directions. Its goal is to ensure that the people in chemical engineering are primed to meet society's needs and rapid changes.

544708 Accelerating Innovation through Academic-Industrial Partnerships

Abstract:

In 2011, The Dow Chemical Company launched the University Partnership Initiative (UPI) to sponsor collaborative research at leading U.S. Universities. This 10-year commitment aims to address fundamental questions in industrial chemistry and chemical engineering with alignment to Dowâ€™s next-generation process technology and product development efforts. To date, this portfolio of projects has generated an impressive array of intellectual property, publications, and numerous examples of the collaborative developments implemented at Dow. This presentation will introduce UPI program and highlight several contributions from our collaborators and leaders in the chemical engineering profession.

540559 Maximizing Uptime, Efficiency, and Safety of Industrial Operations through Early Risk Detection

Abstract:

Achieving zero incidents while maximizing process uptime and efficiency is a key vision for operating companies. An important building block towards this vision is establishment of a proactive risk mitigation culture, supported by effective systems and a strong management. Studies have shown that most of the incidents and unexpected process failures can be avoided if the operating teams get timely information about developing risks and take preventive actions early on. While advances have been made in process monitoring and automation in the past decades, there still remains significant technological and behavioral gaps that prevent plant operations from proactively mitigating process risks.

A modern industrial plant monitors thousands of parameters, generating upwards of 50-100 million data points every day. Thanks to recent breakthroughs in machine learning and artificial intelligence approaches, today there are autonomous systems that can sift through this data and point out meaningful and timely insight. This can help operating teams ascertain process issues that are hidden in the data, long before process variables reach critical levels.

In this presentation, a new autonomous system, Dynamic Risk Analyzer^{TM} (DRA), will be introduced that points out early indicators of process issues using proprietary machine learning. The risk indicators enable operations team to respond and make required process changes, days and sometimes weeks ahead of any alarm, avoiding late stage (expensive) fixes. To realize these objectives fully, a proactive management workflow must also be in place. The presentation will discuss new initiatives and workflows developed by plant operations to implement a proactive culture. Real-life case studies from plants using DRA technology will be presented on how this culture change resulted in increased uptime, efficiency and safety of plant operations.

531309 Gaussian Processes for Hybridizing Analytical & Data-Driven Decision-Making

Abstract:

Surrogate models are widely appreciated in process systems engineering [1]. The typical setting focuses on expensive-to-evaluate, possibly uncertain functions. Examples include: modular process simulators [2], integrated gasification combined cycle processes [3], a carbon capture absorber [4], and many other applications, e.g. [5-8]. Resources are typically limited, so effective decision making requires data-efficient learning.

The data science and statistical machine learning communities typically focus on models learned solely from observed data. But chemical engineering applications may also require explicit, parametric models, e.g. modeling known process constraints, operations constraints, and cost objectives [9]. So, work has integrated semi-algebraic functions with those learned from data [10] or developed semi-physical modeling techniques [11, 12].

This presentation surveys the state-of-the-art in hybridizing analytical and data-driven decision making [13]. We consider three probabilistic modeling applications to these hybrid situations:

Design of experiments for model discrimination [14]. We bridge the gap between classical, analytical methods [15] and Monte Carlo-based approaches [16]. Classical methods may have difficulty managing non-analytical model functions and data-driven Monte Carlo approaches come at a high computational cost. We replace the original, parametric models with probabilistic, non-parametric Gaussian process surrogates learned from model evaluations. The surrogates are flexible regression tools that extend classical analytical results to non-analytical models, while providing us with model prediction confidence bounds and avoiding the computational complexity of Monte-Carlo approaches.

Multi-objective optimization [17, 18]. We make novel extensions to Bayesian multi-objective optimization in the case of one analytical objective function and one black-box, i.e. simulation-based, objective function. The resulting method has been applied to a bone neotissue application [19] and a more general test suite.

Scheduling plant operations under uncertainty. For processes with equipment degradation, we use Gaussian processes to approximate large-scale, mixed-integer optimization problems.

We close by offering a broad outlook on applying probabilistic surrogate models to chemical engineering. Statistical machine learning has recently attracted significant interest in process systems engineering [20]. Here we show that state-of-the-art research in Gaussian processes [21, 22] and probabilistic modeling more generally [23] can have a big impact on chemical engineering.

References

[1] Bhosekar A, Ierapetritou M. Advances in surrogate based modeling, feasibility analysis and and optimization: A

review. *Comput Chem Eng*. 108: 250-267, **2018**.

[2] Caballero JA, Grossmann IE. An algorithm for the use of surrogate models in modular flowsheet optimization. *AIChE J*. 54(10):2633-2650, **2008**.

[3] Lang Y, Zitney SE, Biegler LT. Optimization of IGCC processes with reduced order CFD models. *Comput Chem Eng*. 35(9):1705-1717, **2011**.

[4] Cozad A, Sahinidis NV, Miller DC. Learning surrogate models for simulation-based optim-ization. *AIChE J*. 60(6):2211-2227, **2017**.

[5] Boukouvala F, Ierapetritou M. Surrogate-based optimization of expensive flowsheet modeling for continuous pharmaceutical manufacturing. *J Pharm Innov*. 8: 131, **2013**.

[6] Soepyan FB, Cremaschi S, Sarica C, and others. Estimation of percentiles using the Kriging method for uncertainty propagation. *Comput Chem Eng*. 93:143-59, **2016**.

[7] Shokry A, Ardakani MH, Escudero G, Graells M, Espuña A. Dynamic Kriging-based fault detection and diagnosis approach for nonlinear noisy dynamic processes. *Comput Chem Eng*. 106:758-76, **2017**.

[8] Tran AP, Georgakis C. On the estimation of high-dimensional surrogate models of steady-state of plant-wide processes characteristics. *Comput Chem Eng*. DOI 10.1016/j.comp-chemeng.2018.02.014, **2018**.

[9] Boukouvala F, Hasan MF, Floudas CA. Global optimization of general constrained grey-box models: new method and its application to constrained PDEs for pressure swing adsorption. *J Glob Optim*. 67(1-2): 3-42, **2017**.

[10] Boukouvala F, Floudas CA. ARGONAUT: AlgoRithms for Global Optimization of coNstrAined grey-box compUTational problems. *Optim Lett*. 11(5): 895-913, **2017**.

[11] Pearson RK, Pottmann M. Gray-box identification of block-oriented nonlinear models. *J Process Contr*. 10(4):301-315, **2000**.

[12] Cozad A, Sahinidis NV, Miller DC. A combined first-principles and data-driven approach to model building. *Comput Chem Eng*. 73: 116-27, **2015**.

[13] Boukouvala F, Misener R, Floudas CA. Global optimization advances in mixed-integer nonlinear programming, MINLP, and constrained derivative-free optimization, CDFO. *Eur J Oper Res*. 252: 701-727, **2016**.

[14] Olofsson S, Deisenroth MP, Misener R. Design of Experiments for Model Discrimination Hybridising Analytical and Data-Driven Approaches. *arXiv:1802.04170*. **2018**.

[15] Buzzi-Ferraris G, Forzatti P, Emig G, Hofmann H. Sequential experimental design for model discrimination in the case of multiple responses. *Chem Eng Sci*. 39(1): 81-85, **1984**.

[16] Vanlier J, Tiemann CA, Hilbers PAJ, van Riel NAW. Optimal experiment design for model selection in biochemical networks. *BMC Syst Biol*, 8(20), **2014**.

[17] Olofsson S, Mehrian M, Geris L, Calandra R, Deisenroth MP, Misener R. Bayesian Multi-Objective Optimisation of Neotissue Growth in a Perfusion Bioreactor Set-Up. *Computer Aided Chemical Engineering*. 40: 2155-2160, **2017**.

[18] Beykal B, Boukouvala F, Floudas CA, Pistikopoulos EN. Optimal design of energy systems using constrained grey-box multi-objective optimization. *Comput Chem Eng*. DOI 10.1016/j.compchemeng.2018.02.017, **2018**.

[19] Mehrian M, Guyot Y, Papantoniou I, Olofsson S, Sonnaert M, Misener R, Geris L. Maximizing neotissue growth kinetics in a perfusion bioreactor: An *in silico* strategy using model reduction and Bayesian optimization. *Biotechnol Bioeng*. 115(3):617-29, **2018**.

[20] Lee JH, Shin J, Realff MJ. Machine learning: Overview of the recent progresses and implications for the process systems engineering field. *Comput Chem Eng*. DOI 10.1016/j.compchemeng.2017.10.008, **2017**.

[21] Damianou A, Lawrence N. Deep Gaussian processes. In *Artificial Intelligence and Statistics*. 207-215, **2013**.

[22] Deisenroth MP, Ng JW. Distributed Gaussian processes. In *32nd International Conference on International Conference on Machine Learning*. 37: 1481-1490, **2015**.

[23] Ghahramani Z. Bayesian non-parametrics and the probabilistic approach to modelling. *Phil Trans R Soc A*. 371(1984):20110553, **2013**.

534888 Award Submission: Overcome Drug Resistance of Cancer Cells By Confining, Perturbing and Analyzing Them in Nano-Liter Chambers One Cell at a Time

Abstract:

Tumor cells develop resistance to many therapies. And it is a major problem across many non-surgical cancers. We find that, multi-functional nanotechnology, when applied onto drug resistance development scenario, could uncover the resistance mechanism and provide rationale for designing effective combination therapy to overcome it.

Although it may take a few months to see the reoccurrence of tumor, yet people recently realize that tumor cells already developed resistance as early as a few days after drug treatment. This is because that tumor cells can easily change its cell state from the original drug sensitive one to a very different one, which allow them to survive under the drug treatment. Therefore, stop the unfavorable cell state transition is required in order to overcome drug resistance. To do that, it is essential to discover the driver regulators that lead the unfavorable cell state changes. Finding the driver regulators are very hard because of 2 major obstacles. Because there are mixtures of different cell states involved in such transition (need to analyze individual cells), and because, within each cell, the regulatory network has many regulators involved (need to characterize many proteins), single-cell proteomic analysis is the best way to study it.

However, single-cell proteomics are extremely hard to implement mainly because the protein amounts in single cells are too little to be detected. Fortunately, this problem could be beautifully addressed by a multi-functional bionanotechnology developed in our lab called single-cell barcode chip (SCBC). This nanotechnology can confine individual cells in nano-liter chamber. Perturbing and lysing single cell within the chamber is also feasible through programming the nano-liter fluid. Consider the nano-liter size of each chamber, small amount of protein from single cell when released within such tiny volume will also have a detectable concentration. Based on such multi-functional, programmable, nano-chamber, many important regulators from each individual cell are quantified simultaneously with the help of arrays of antibodies and nano-probes installed orderly within each chamber.

When applied to tumor cells that are under therapeutic treatment, this powerful nanotechnology enables us to uncover the changes of many important regulators simultaneously across thousands of individual cancer cells. The big data generated from the nanotechnology pinpoint the most important regulators that are driving the initiation of such unfavorable cell state changes, which leads to resistance. Base on the information from the powerful nanotechnology, an effective combination therapy is designed which is experimentally validated to be able to halt the unfavorable cell-state transition and therefore arrested the resistance development.

This powerful nanotechnology could provide deep biological insights that are otherwise impossible to see. The nanotech has the potential to uncover the resistance mechanism across many other cancer types and potentially bring durable therapy response to a broad range of cancer patients. This may usher a new era of nanotech-assisted rational designing of combination therapy.

538040 Award Submission: Chitosan / Cellulose Nanocrystals / Calcium Phosphate Hydrogels for Vertebral Compression Fracture Treatment

Abstract:

Vertebral compression fractures are the most common consequence of osteoporosis accounting for approximately 700,000 out of the 1.5 million total osteoporotic fractures that occur annually in the USA. Injection of poly(methyl methacrylate), a non-biodegradable and non-bioactive polymer, into the vertebral bode as a mechanically stiff bone cement is the most common treatment for vertebral compression fractures. Recently there has been a growing interest to use bioactive, degradable, and mechanically suitable substitutes for the treatment of these fractures. In this research, we focused on creating injectable, chitosan-based hydrogels for the treatment of vertebra compression fractures. First, we investigated the effect of different ratios of ionic and covalent crosslinking on hydrogel physical stability and mechanical properties. Then, cellulose nanocrystals with varying surface charge were utilized to modify the mechanical properties and physical stability of the chitosan matrix. Finally, different calcium phosphates with various solubility were incorporated into the matrix in order to achieve sustained release of calcium and phosphate ions for optimal bone inductivity outcomes without inducing unwanted cytotoxicity.

535833 Award Submission: The Binary Effect on Drug-Resistant Bacteria of Polymeric Vesicles Appended By Proline-Rich Amino Acid Sequences and Inorganic Nanoparticles

Abstract:

Prevalent research underscores efforts to engineer highly sophisticated nano-vesicles that are functionalized to combat antibiotic-resistant bacterial infections, especially those caused by methicillin-resistant *Staphylococcus aureus* (MRSA), and that aid with wound healing or immunomodulation. This is especially relevant for patients who are susceptible to *S. aureus* infections post-operatively. Technical challenges associated with this aim require a thorough assessment of the chemical properties of synthesis materials and methods. Here, formulations were incorporated into polymeric, biocompatible vesicles called polymersomes that self-assemble via hydrophobicity interactions of admixed aqueous and organic substances. Nano-polymersomes were synthesized using a high molecular weight amphiphilic block copolymer, and were conjugated to include antimicrobial peptides (AMPs) along the peripheral hydrophilic region and silver (Ag) nanoparticles inside their hydrophobic corona. *In vitro* testing on bacterial and human cell lines indicated that finely tuned treatment concentrations of AMP and Ag nanoparticles in polymersomes synergistically inhibited the growth of MRSA without posing significant side effects, as compared with other potent treatment strategies. In particular, a ratio of silver-to-AMP of about 1:1.9, corresponding to approximately 0.1 mg/ml of silver nanoparticles and 40 Î¼M of the peptide, yielded complete MRSA inhibition over a 23-hour time frame. The silver nanoparticles generate bacterial cell wall indentations that prompt the influx of external entities and the outflow of vital proteins or nutrients. This weakening of lipopolysaccharide permeability barriers promotes the unimpeded cellular translocation of contiguous AMP molecules, which can then disrupt internal processes including DNA or protein synthesis. This bacteriostatic activity, coupled with nominal cytotoxicity towards native human dermal fibroblast cells, extends the potential for AMP/Ag nanoparticle polymersome therapies to replace antibiotics in the clinical setting. This study will facilitate contemporary research that is feasible, important, and highly relevant, and possesses the potential to advance knowledge in several fields, including drug delivery, nanotechnology, and artificial immunity.

522740 Award Submission:Â Lipid Nanoparticle Ionization at Endosomal pH Is a Cell-Free Predictor of mRNA Delivery Efficacy In Vivo

Abstract:

Messenger RNA (mRNA) therapy has the potential to transform the lives of millions of people suffering from a wide range of diseases by augmenting protein expression or replacing missing or mutated proteins. The versatility of mRNA-based drugs has made it a leading gene therapy platform, with dozens of studies utilizing it for applications such as such as protein replacement, prophylactic vaccines, cancer immunotherapy, gene editing, and cellular reprogramming. However, the clinical translation of mRNA therapeutics is contingent upon their safe and effective delivery into target cells. Lipid nanoparticles (LNPs) have emerged as a promising class of delivery materials with the potential to facilitate mRNA delivery. Given the high cost of preclinical testing, most delivery systems are screened in cell culture before advancing into murine and non-human primate models. Unfortunately, cell culture testing is resource intensive and prone to high rates of false negatives and positives. Here, we describe an alternative, cell-free screening strategy that accurately predicts mRNA delivery potency in animals. This was accomplished through the synthesis of a small library of lipid-like (lipidoid) materials, a subset of which effectively delivered mRNA to the liver and spleen of mice following systemic injection of mRNA-LNPs at a 0.75 mg/kg dose. In particular, the lipidoid $306O_{i10}$, which contains a branched tail, facilitated marked levels of firefly luciferase expression in mice. A structure-function analysis using our *in vivo* results revealed that *in vitro* efficacy, particle size, mRNA entrapment, and pKa did not correlate with efficacy in mice. However, the degree of lipid nanoparticle surface ionization at the late endosomal pH of 5.0 was a strong predictor of *in vivo* efficacy (R^2= 0.95). The generation and testing of a secondary library of 33 lipidoids confirmed LNP surface ionization as an effective screening tool, correctly predicting materials with both high and low efficacy in mice. These data indicate that the measurement of nanoparticle surface ionization is an inexpensive, high throughput, and reliable method to identify mRNA delivery materials that are highly potent *in vivo*.

535666 Award Submission:Â Nanomagnetic Illuminators for In Vivo Optical Imaging of Osteoarthritic Knee Joints

Abstract:

Nanoparticle formulations of various sizes and charges are being developed to diagnose and treat osteoarthritis (OA), a maladaptive disease of the joint. However, the transport and *in vivo* bio-distribution of nanocarriers in the joint space remain poorly understood. Developing a biocompatible platform for visualizing and quantifying the distribution of nanoparticles in the joint space could help engineer better nanoparticles for drug delivery and disease diagnosis. Here, we focus on block copolymer stabilized nanomagnetic illuminators obtained by encapsulating near-IR lipophilic dyes. We apply both in-vivo imaging system (IVIS) and electron paramagnetic resonance spectroscopy (EPR) to visualize and quantify the distribution of these multifunctional magnetic carriers in the healthy and osteoarthritic joints of Lewis rats.

Magnetic nanoparticles synthesized by a high temperature synthesis method under aerobic conditions, block copolymer PEG_{5kDa}_$PDLLA_{6kDa}$, and lipophilic tracer 1,1'-Dioctadecyl-3,3,3',3'-tetramethylindotricarbocyanine iodide (DiR) were dissolved in the water miscible solvent THF. Rapid mixing of the solutes against water (anti-solvent) using a confined impingement jet mixer under high local supersaturation condition resulted in kinetically frozen structures of about 200 nm. The magnetic composite nanocarriers were purified via magnetic separation and dye encapsulation was confirmed using absorbance measurements and an *in vivo* imaging system.

Toxicity of the DiR loaded magnetic nanocarrier on fibroblast-like synoviocytes was tested using an MTS assay, a colorimetric assay used to analyse cell metabolism. Greater than 80% of the synoviocytes were found viable after 24 hours of particle exposure in the concentration range of 0.1 to 0.8 mg iron oxide/mL, suggesting negligible toxic effects of the nanocarrier under these conditions.

Biodistribution and particle retention studies were performed on 15-month old Lewis rats with healthy and OA induced knees. 60 ug of DiR loaded magnetic composite nanocarrier was injected and distribution of the nanomagnetic illuminator was tracked for 14 days. Further, the retention of the particles in organs of euthanized rats was visualized by IVIS imaging and quantified via EPR. IVIS images of the organs analyzed for radiant efficiency suggested that particles were predominantly retained in the knee joint and cleared faster in a healthy joint as compared to an osteoarthrtic knee. Based on the EPR measurements on the lyophilized organs, we found that ~60 % of the injected iron oxide nanoparticles remained in the joint space. Our studies involving both visualization and quantification of DiR loaded magnetic particle indicate that this carrier can be explored as a safe option to better understand transport of nanoparticles in tissues of joint space and also to deliver hydrophobic drugs for treatment.

537428 Award Submission: Quantification of Inflammatory Response and Morphological Change of SIM-A9 Microglia By Neuro-Probes

Abstract:

Fluorescent probes such as calcium dyes,[1] neurotransmitter nanosensors,[2] and voltage probes,[3] are utilized ubiquitously in neuroscience to study brain structure and neurochemical activity. While most of these probes target the study of neurons, their effects on another major cell type of the brain, microglia, are often overlooked. Microglia are the resident immune cells of the brain that number in the billions, and are central in maintaining the homeostasis of the neuronal environment through states of activation and quiescence.[4] In the brain, microglia mediate inflammatory responses to foreign and pathogenic material, which can elicit downstream neurotoxicity and neurodegeneration. Herein, we present a study on the effects of several brain chemistry probes on a SIM-A9 mouse microglial cell line. We show that single-walled carbon nanotube (SWNT)-based dopamine nanosensors are uptaken by, but are not cytotoxic to, SIM-A9 mouse microglia. The activity of SIM-A9 mouse microglia is next quantified with quantitative PCR and with an immune proteomics screen to monitor mRNA and protein expression and secretion levels of common inflammatory cytokine biomarkers. We show that SIM-A9 co-incubation with dopamine nanosensors induces a slight increase in expression of inflammatory cytokines TNF-α, iNOS, IL-1β, and IL-6. The observed nanosensor-induced cytokine upregulation is significantly lower than that caused by lipopolysaccharide stimulation. Live-cell imaging results further reveal that SIM-A9 interactions with SWNT nanosensors elicit a significant change in SIM-A9 cell morphology towards an elongated and ramified state, an *in vitro* indicator of a quiescent rather than inflammatory microglial state that halts cell proliferation. These contrasting effects indicate that nanosensors affect multiple microglial signaling pathways, possibly through a non-canonical immune response. We also discuss how common passivation techniques, namely PEGylation, can mitigate microglial inflammatory responses to brain imaging probes. PEGylated nanosensors show lower uptake into SIM-A9 cells, a reduced inflammatory response, and partial mitigation of SIM-A9 morphological change. Our results motivate greater attention to the effects of ubiquitously utilized neuro-imaging probes on the brain microglia.

References

1. Grienberger, C. & Konnerth, A. Imaging Calcium in Neurons. *Neuron* **73,** 862–885 (2012).
2. Beyene, A. G., McFarlane, I. R., Pinals, R. L. & Landry, M. P. Stochastic Simulation of Dopamine Neuromodulation for Implementation of Fluorescent Neurochemical Probes in the Striatal Extracellular Space. *ACS Chem. Neurosci.* **8,** 2275–2289 (2017).
3. Peterka, D. S., Takahashi, H. & Yuste, R. Imaging Voltage in Neurons. *Neuron* **69,** 9–21 (2011).
4. Kierdorf, K. & Prinz, M. Factors regulating microglia activation. *Front. Cell. Neurosci.* **7,** 1–8 (2013).

530293 Award Submission: Cardiac Troponin I Detection Using Antibody-Immobilized Disposable Cover Glass and AlGaN/GaN High Electron Mobility Transistors

Abstract:

Cardiac troponin I (cTnI) detection was demonstrated using a functionalized disposable cover glass connected to an AlGaN/GaN high electron mobility transistors (HEMT). Cardiovascular disease is one of the leading causes of mortality in the US, and it has caused tremendous hardship for both individual health and finance in United States. Every year, over 800,000 Americans die due to the cardiovascular diseases.[1] Troponins are released when the cardiac and skeletal muscles are damaged. There are three types of troponins; troponin I, troponin T, and troponin C. . Troponin I is extremely cardio-specific, and the elevation of troponin I concentration in the blood stream indicates the cardiac muscle damage. Troponin T is also a biomarker for myocardial infarction patient, however it is not exclusive. Studies have shown the elevated troponin T were also detected for patients with muscular dystrophy and renal disease. With the non-cardio specific nature of troponin C, it has never been used for clinical practice as a myocardial necrosis marker. [2,3,4] Currently, all the detection methods and systems for cTnI are time consuming, expensive and require a well-trained technician to perform and complete the tests.

The challenge for cardiac Troponin I (cTnI) detection method in clinical practice is to develop a hand-held, low cost, accurate, and real-time biosensors for myocardial infarction patients. The combination of functionalized cover glass and AlGaN/GaN high electron mobility transistors (HEMT) detection method has overcome the primary difficulty of biomarker solution with high ionic strength in traditional HEMT biosensors. In this approach, the HEMT device is not exposed to the bio-solution, which can be used repeatedly. The sensor functionalization was performed on a low cost disposable glass slide, which can be viewed as testing strips. Furthermore, the cover glass could be replaced by plastic or paper for cTnI detection, which can be more cost-effective. To test the cTnI concentration, pulsed bias-voltages are applied to a functionalized electrode on the cover glass and the drain terminal of the HEMT device. The electrical pulse would result in the formation of the electrical double layers (EDL) on the active sensing area, and different cTnI concentration would lead to different charge potential that is applied onto the conducting channel of this HEMT device. The detected cTnI concentration range was achieved from 0.1 ng/ml to 100 ng/ml. This covers wide range of cTnI concentration for the myocardial infraction, such as the elevated cTnI concentration of 10ng/ml or higher for acute myocardial infraction (AMI), and 0.5 ng/ml to 2.0 ng/ml cTnI concentration for early stage of AMI patients. The static drain current changes as a function of cTnI concentrations were modeled with the Langmuir extension model. Also, the spring relaxation model was employed for the dynamic modeling of the time-dependent current response for different biomarker concentrations. The results in this study demonstrated that the functionalized cover glass approach is a very attractive cTnI detection technology for inexpensive cartridge-type heart attack biosensors, and this technology can be applied to other sensing areas such as cancer detection, environmental monitoring, and food safety.

1. Benjamin, E.J., Blaha, M.J., Chiuve, S.E., Cushman, M., Das, S.R., Deo, R., de Ferranti, S.D., Floyd, J., Fornage, M., Gillespie, C., Isasi, C.R., JimÃ©nez, M.C., Jordan, L.C., Judd, S.E., Lackland, D., Lichtman, J.H., Lisabeth, L., Liu, S., Longenecker, C.T., Mackey, R.H., Matsushita, K., Mozaffarian, D., Mussolino, M.E., Nasir, K., Neumar, R.W., Palaniappan, L., Pandey, D.K., Thiagarajan, R.R., Reeves, M.J., Ritchey, M., Rodriguez, C.J., Roth, G.A., Rosamond, W.D., Sasson, C., Towfighi, A., Tsao, C.W., Turner, M.B., Virani, S.S., Voeks, J.H., Willey, J.Z., Wilkins, J.T., Wu, J.H., Alger, H.M., Wong, S.S., Muntner, P., 2017. Heart disease and stroke statisticsâ€"2017 update: a report From the American Heart Association. Circulation (2017)
2. Mythili and N. Malathi, Diagnostic markers of acute myocardial infarction. Biomed. Rep. 3, 743 (2015).
3. Amitava Dasgupta and Amer Wahed, Chapter 8 - Cardiac Markers, In Clinical Chemistry, Immunology and Laboratory Quality Control, Elsevier, San Diego, 2014, Pages 127-144, ISBN 9780124078215
4. Allan S. Jaffe, Vlad C. Vasile, Margherita Milone, Amy K. Saenger, Kalen N. Olson, Fred S. Apple, Diseased Skeletal Muscle: A Noncardiac Source of Increased Circulating Concentrations of Cardiac Troponin T, Journal of the American College of Cardiology, Volume 58, Issue 17, Pages 1819-1824 (2011)

528485 Ceria Nanoparticle Dissolution and Stability in Acidic Aqueous Environments

Abstract:

Ceria (cerium oxide) nanomaterials, or nanoceria, have many applications such as acting as a redox catalysis and in catalyst supports, solid oxide fuel cells, sintering additives, heat resistant alloy coatings, and as UV absorbents and superconducting materials. The cerium atoms on the surface of nanoceria can store or release oxygen, cycling between Ce^{3+} and Ce^{4+}, and can therefore relieve oxidative stress within the body. As an anti-oxidant, nanoceria has to potential to reduce nausea and bacterial reproduction, stimulate the immune system to assist in fighting infection and disease, and prevent cancerous tumor growth. Nanoceria dissolution is present in acidic environments *in vivo*. In order to accurately define the fate of nanoceria *in vivo*, nanoceria dissolution or stabilization is observed *in vitro* using acidic aqueous environments.

Nanoceria stabilization is a known problem even during its synthesis; in fact, a carboxylic acid, citric acid, is used in many synthesis protocols. Citric acid adsorbs onto nanoceria surfaces, capping particle formation and creating stable dispersions with extended shelf lives. For example, citric acid surface coatings prolong nanoceriaâ€™s residence time within the blood stream. Dynamic light scattering (DLS) was utilized to determine the hydrodynamic diameter of nanoceria in aqueous solutions as a measure of dispersion stability. Nanoceria is shown to agglomerate in the presence of some carboxylic acids over a time scale of up to 28 weeks, and degrade in others, at pH of 4.5 (representing that of phagolysosomes). Sixteen carboxylic acids were tested: citric, glutaric, tricarballylic, 2-hydroxybutyric, 3-hydroxybutyric, adipic, malic, acetic, pimelic, succinic, lactic, tartronic, isocitric, tartaric, dihydroxymalonic, and glyceric acid. Each acid is introduced as 0.11M, into pH 4.5 iso-osmotic solutions. Controls such as ammonium nitrate, sodium nitrate, and water are also tested to assess their effects on nanoceria dissolution and stabilization.

To further test stability, nanoceria suspensions are subjected to light versus dark milieu. Light induced nanoceria agglomeration in some, but not all ligands, and is likely to be a result of UV irradiation. Light initiates free radicals generated from the ceria nanoparticles. Some of the ligands completely dissolved the nanoceria when exposed to light. Citric, malic, and lactic acids form coordination complexes with cerium on the surface of the ceria nanoparticle that can inhibit agglomeration. This approach identifies key functional groups required to prevent nanoceria agglomeration. The impact of each ligand on nanoceria are analyzed and will ultimately describe the fate of nanoceria *in vivo*.

Supported by NIH Grant 1R01GM109195.

525648 Probing Peptoid-Carbon Nanotube Coatings for Biological Imaging

Abstract:

Single-walled carbon nanotubes (SWNT) have emerged as ideal candidates for biological imaging due to their tissue-transparent near-infrared fluorescence, and infinite photobleaching lifetimes.[1] However, the hydrophobic surfaces of SWNT require ligand functionalization or polymer encapsulation to promote aqueous colloidal stability for use in biological environments. We introduce peptoids, N-substituted glycine polymers, as easily synthesized programmable polymers with a large monomer space and that are less susceptible to proteolytic degradation than peptide or protein polymers.[2] We show that peptoids can be used as a novel carbon nanotube coating by pinning peptoid polymers to the SWNT surface through pi-pi and electrostatic interactions. Herein, we identify the stabilizing interactions of peptoids on SWNT surfaces. We probe the effect of peptoid polymer length, hydrophilicity, initial dispersion conditions, and ionic buffer conditions on the resulting colloidal stability of peptoid-SWNT conjugates. These findings allow us to stabilize carbon nanotubes with rationally designed peptoid polymers and enable us to effectively use peptoid-based SWNT coatings for biological imaging.

1. Zhang, J. *et al.* Molecular recognition using corona phase complexes made of synthetic polymers adsorbed on carbon nanotubes. *Nat. Nanotechnol.* **8,** 959â€"68 (2013).
2. Nam, K. T. *et al.* Free-floating ultrathin two-dimensional crystals from sequence-specific peptoid polymers. *Nat. Mater.* **9,** 454â€"60 (2010).

529604 Synthesis, Characterization, and Interfacial Properties of Lignin Coated Iron Oxide Magnetic Nanoparticles in Aqueous Solutions

Abstract:

Lignin is an abundant natural raw material on the planet with attractive properties, including high thermal stability, adsorbent properties, and antioxidant characteristics. However, lignin is a waste product with very low-value applications in the paper pulp industry. Ongoing studies are focusing on the development of value-added high-performance materials using lignin, which includes the development of nano- and micron-sized particles. In this study, core-shell nanostructures were synthesized using magnetic nanoparticles as the core functionalized by lignin macromolecules as the shell. In addition to the versatile properties of lignin, the hybrid nanostructure is attractive due to their ease of manipulation under the presence of a magnetic field and their inherent superparamagnetic properties. In this work, magnetic iron oxide-lignin nanoparticles were synthesized using two methods: a co-precipitation technique using iron salts to produce core-shell nanostructures (lignin@Fe_3O_4), and nanoprecipitation using ethanol/water solutions. For the co-precipitation technique, lignin was dissolved in ammonium hydroxide; whereas a lignin solution in ethanol was used for the nanoprecipitation method. Various concentrations of lignin were used to stabilize iron oxide nanoparticles, and the stabilization or the aggregation of the nanoparticles were characterized as a function of the weight ratios between lignin and the magnetic core for both synthetic routes. The nanoparticles were characterized by using a variety of experimental techniques including DLS, TEM, UV-Vis, FTIR, XRD, VSM and zeta potential analyses. The hybrid nanoparticles showed various sizes ranging from 17 -25 nm for single core-shell lignin@Fe_3O_4 nanostructures and from 100-200 nm for the aggregates. TEM images of the particles displayed encapsulation of iron oxide cluster within the lignin polymer matrixes at high concentrations of lignin. The surface tension of lignin@Fe_30_4 nanoparticles in water and ethanol/water mixtures was measured at different nanoparticle concentrations using the pendant drop method. The results of this work show that superparamagnetic iron oxide nanoparticles coated with lignin were successfully synthesized. The outstanding properties of lignin as a renewable material, which could be used as antioxidant, emulsion stabilizer, UV adsorbent, antimicrobial properties, and preparation of biomaterials make the lignin-magnetic nanocomposite a unique substrate for a plethora of future investigations.

530075 Investigation of Interactions between Magnesium Silicate Particles and Diamond-like Carbon Surface By Atomic Force Microscopy

Abstract:

Contamination of nanoparticles on surfaces becomes a growing concern in various industries, especially in microelectronic industries since the size of the components have drastically reduced. Although the atmospheric dust is generally controlled, particles from various sources such as processing components and operating environments are often detected. This work emphasizes on the adhesion of magnesium silicate particles onto diamond-like carbon (DLC) surface. Magnesium silicate particles are applied in many daily uses such as skin-care products, cosmetics, or even papers, while DLC thin film has been considered as a promising material for protection of electronic components because of its high hardness, low friction, high thermal conductivity and chemical inertness. Moreover, due to low hardness of magnesium silicate that leads to strong adhesion to the surface, removal of magnesium silicate particles becomes a challenging problem.

In this work, two types of magnesium silicate, i.e., commercial talcum powder and Mg_2SiO_4 particles synthesized by the hydrothermal method, were studied. The magnesium silicate-DLC interaction was investigated by atomic force microscopy (AFM) operated in force-volume mode, in which DLC coated silicon AFM tip was used. The particles were deposited on a silicon wafer, washed with deionized water, and dried at room temperature before subjected to the AFM analysis.

Preliminary experiments, which compare AFM analyses of talcum particles using a bare silicon tip to that using a DLC coated silicon tip, revealed that the adhesion forces measured are significantly different, although the spring constants of both tips are similar. The average adhesion forces measured using the bare silicon tip and the DLC coated silicon tip are 80 nN and 40 nN, respectively. Hence, tip-particle interaction can be used to represent particle-surface interaction.

In the experiments comparing talcum and Mg_2SiO_4 particles, the normal spring constants of the DLC coated tips were calibrated by thermal noise method (k = 0.02 N/m). The adhesion measurements were conducted several times to achieve statistical representation of the data. According to the measured force-displacement curves, both adhesion force and force acting on both types of particles are significantly different. The average adhesion force on talcum powder tends to be larger than that on Mg_2SiO_4 particles. Although the difference is suggested to be a result from different hardness of the particles (i.e., the hardness of talcum is 1-1.5 in Mohsâ€™ scale, while that of Mg_2SiO_4 is 7), other factors such as chemical composition may also involve.

To verify the effect of chemical functional group on the surface of magnesium silicate particle on the tip-particle interaction, surface modification was conducted. Talcum naturally contains surface hydroxyl, but the content of the hydroxyl group can be varied by a treatment with various kinds of acid. For surface modification, 100 g of talcum powder was immersed in 1 L of concentrated acid at 80°C for 2 h. The acids employed were formic acid, acetic acid and hydrochloric acid. The treated particles were filtered, washed with deionized water and dried in an oven at 70°C for 48 h. According to Fourier transform infrared spectroscopic analyses, the talcum powder treated with hydrochloric acid showed significant increase in the signal corresponding to hydroxyl group, while that of the powders treated with organic acids is roughly unchanged. However, the adhesion force of hydrochloric acid-treated talcum particle is about the same as the untreated talcum. On the other hand, the talcum treated with organic acids showed much increased adhesion forces. It is therefore suggested that hydroxyl group does not contribute to magnesium silicate-DLC interaction. Instead, grafting of the surface of talcum by organic groups, which leads it to be hydrophobic, can enhance the adhesion because DLC is also hydrophobic. Hence, hydrophobicity of the particles plays an important role in adhesion onto the DLC surface in addition to the softness of the particles.

533086 Polymer-Metal Composite Nanoparticles Via Vapor Phase Deposition Processes Onto Liquid Substrates

Abstract:

The tuning of polymer nanoparticles is important for applications in catalysis, optics, and electronics. Metal nanoparticles have already shown promise in these applications because of their high surface area-to-volume ratio and durability. However, methods to create polymer-metal nanocomposites are expensive, time consuming, and require the use of many solvents. In this work, we combine the vapor deposition methods of sputter deposition and initiated chemical vapor deposition (iCVD) to create metal-polymer composite nanoparticles. By combining these two vapor deposition methods, we are able to tune the polymer-metal composite functionality independently. We characterize the resulting structures using dynamic light scattering and transmission electron microscopy.

535945 Controlling Surface Morphology and Spatial Distribution of Active Nanoinclusions in Functional Coatings Via Air-Controlled Electrospray Process

Abstract:

Electrospray process is used to produce electrically charged droplets of sizes in micron and submicron range. This process is used in diverse applications like production of thin and uniform coating, nanodroplet and nanostructure production, biomedical engineering, energy storage and mass spectrometry. In this process, the electrospraying liquid is pumped out of a needle through a syringe and an electric field is applied between the needle and the collecting plate. When the liquid is released out of the needle the electric field causes the charges to separate inside the liquid meniscus and it takes the shape of a cone at the needle, which is called as Taylor cone. When Taylor cone becomes unstable, depending on the process conditions either jets or droplets are ejected from the cone. In the spray process for viscoelastic liquids, the jet extends due to viscoelastic forces and the modulations on the jet radius due to axisymmetric instabilities cause the jet to break-up. Whereas, in non-viscoelastic liquids, the atomization of the meniscus at the nozzle is responsible for the formation of droplets. In the electrospray process, the electric field force is responsible for the growth of the axisymmetric instability and breaking the jet into droplets. In the air spray process, the external air flow is applied axially to the jet and the air drag accelerates the growth of the axisymmetric instability. In this work, we study the spray phenomena with simulations and experiments for the following cases: i) electrospray (only electric field), ii) air spray (only airflow) and iii) air controlled electrospray (airflow and electric field).

In this work, we use a Lagrangian discretized model as it is suitable to model both the jet breakup and droplet dynamics. In the discretized model, we consider the liquid jet to be made up of series of beads attached together with springs. The Newtonâ€™s second law of motion is applied to conserve the momentum. The jet at the nozzle is perturbed by introducing a normal mode disturbance to the radius of the bead at the nozzle. This disturbance results in axisymmetric instability on the jet and with the growth of the instability the jet breaks into droplets. We incorporate equations of motion for the formed droplets to study their dynamics. In this model, we also model the spray process due to external air flow by incorporating the air drag effects for air spray and air controlled electrospray cases.

In addition, we perform experiments to study the air controlled electrospray process of 8 wt% polyvinylidene fluoride in dimethylformamide solution. The air spray, electrospray, and air controlled electrospray processes are visualized with a high-speed camera. The average radius of the droplets obtained from experiments is compared with simulation results. The surface roughness and efficiency of the coating of PVDF/DMF solution on the collector are obtained to study the effect of air flowrate on air controlled electrospray of the solution. We also investigate the effect of air flow in air controlled electrospray process of 3% PVA/H2O solution with 15 vol.% Carbon Black or Carbon Nanotube particles to produce coatings with precisely controllable nanoscale topology, morphology, deposition thickness, and spatial distribution of active nano-inclusions. We further investigated the effect of additional deformation on the deposition homogeneity and the dispersion of carbon NPs (Carbon Black and Carbon Nanotubes) in resulting coatings, as it plays a crucial role in energy-storage applications.

541647 Can Soft-Matter Mechanics Provide New Avenues for Remediating (and even preventing!) Biofilm Infections?

Abstract:

Biofilms are systems of microbes that are bound to each other, and often to an external material, by a matrix of biopolymers and proteins. Therefore, biofilms are characterized by mechanical properties that are not present for systems of equivalent microbes in the non-biofilm state, with individual organisms suspended in fluid. We investigate how these mechanical properties contribute to the pathogenesis of biofilms, and how mechanical properties of the biofilm and/or its environment might be altered to make biofilms easier to clear, and even to prevent biofilms from developing.

We recently found that chronic, years-long biofilm infections evolve to change the production of matrix materials in such a way as to increase or maintain the mechanical robustness of the biofilm; we use shear rheology to measure mechanical robustness in terms of elasticity, yield strain and stress, and the energy input required to cause the biofilm to yield. More recently, we have begun investigations of how these mechanical properties of biofilms could impact the ability of the immune system to clear biofilms. Our preliminary results show that varying the elasticity of abiotic gels over the range measured for clinical biofilms has greatly influences how likely it is that neutrophils (phagocytic immune cells) will break off and engulf parts of the gel. This suggests that weakening relevant mechanical characteristics of biofilms may help to promote their clearance by neutrophils, which are the body's first responders and the primary defenders against neutrophils. We will present early results on approaches to disrupt biofilm mechanics by attacking the interactions between specific matrix materials.

541736 Mechanisms of Bacterial Biofilm Growth and Biofilm-Virus Interactions

Abstract:

On most moist surfaces, bacteria grow into communities termed biofilms, which display strongly increased resistance to antibiotics, and constitute the most abundant form of biomass on Earth. I will first present new microscopy and imaging techniques we recently developed to follow dynamical processes in biofilms at the single-cell level. I will then proceed to show how we applied this novel imaging technology in combination with genetics and molecular techniques to discover mechanisms of basic processes in biofilm development: I will show how individual cells within biofilms interact with each other, how cells disperse from biofilms, and how bacterial biofilms interact with bacteriophage viruses, which are the most frequently encountered predators of bacteria in the environment. This work on biofilm dynamics at the single cell level promises many new discoveries for preventing biofilm growth on environmental and biomedical interfaces.

541645 The Biophysics of Bacterial Biofilms Facilitate Surface Survival in Moving Fluids but May Reveal an Achilles Heel

Abstract:

Bacterial biofilms are microscopic assemblages of bacterial cells usually attached to a surface and held together by a self-produced extracellular polymeric slime (EPS) matrix. Biofilms are ubiquitous in the natural environment and are highly problematic in industry and medicine where they cause corrosion, fouling, contamination and chronic medical and dental infections. The EPS matrix is chemically complex and is composed of polysaccharides, lipids, proteins and DNA and protects the bacteria within from antibiotics chemical challenges and host immunity. Mass transfer through the EPS is dominated by diffusion which, allows the development of gradients in nutrients, waste products and cell signals. The bulk mechanical properties of biofilms play an important role in survival by allowing the biofilm to respond to fluid shear stresses over very short (ms) and very long (days to weeks) time scales. Creep and relaxation tests show that biofilms behave as viscoelastic liquids however and recent observations of high velocity impacts with water droplets suggest they rapidly form interfacial instabilities allowing them to flow over surfaces with velocities of meters per second. Generating turbulence in the biofilm might explain enhanced killing by antimicrobial agents in dental biofilms and have broader applications in biofilm control. A better understanding of how bacterial biofilms respond to fluid flow provides new opportunities to develop more effective control strategies.

541646 Biofunctionalization of Implants through Thin Films

Abstract:

Control of the tissue-implant interface is integral to the success of the implant, and ultimately the health of the patient. Through a series of projects on multiple types of implant substrates, the Gawalt Lab has developed a system for immobilizing multiple bioactive molecules on the surface to control the interface using strongly adhered linkers and basic organic chemistry. For example, ceramic surfaces have been modified as bone fillers to enhance bone growth and reduce infection with either one or two molecules, while stent materials are being modified to reduce platelet aggregation and stimulate endothelization. Recently, a triple-functionalized surface has been completed on orthopedic implant materials to simultaneously enhance bone growth, while reducing bacterial adhesion. This new system is the first system with three bioactive molecules in a coating.

541773 Infection-Resisting Biomaterials

Abstract:

The synthetic surfaces associated with tissue-contacting biomedical devices are attractive to many bacteria, because these surfaces provide anchoring sites, often well protected from the innate immune response, in a moist environment with abundant access to nutrients. Clinically, bacterial colonization of device surfaces is significant, because a series of revision surgeries is often required to resolve the resulting implant-associated infection. The impact on both the patient and the health-care system can be substantial. While better surgical practice can mitigate the chances of infection, eliminating bacterial surface colonization is in large measure an engineering problem. One solution has been to incorporate antimicrobials into the device. These then elute into the surrounding physiological system over extended time periods after surgery. This approach, while FDA-approved in a small number of applications such as antibiotic-loaded bone cement, delivers antibiotic whether it is needed or not, and it thus runs counter to the rapidly increasing concerns about the overuse of antibiotics and the ongoing evolution of antibiotic-resistant bacterial strains. An alternate concept is that of self-defensive surfaces. These sequester antimicrobials until their release in response to a bacterial trigger. Sequestered antimicrobials are present only in very small amounts, and they are released only if needed. Consequently, they do not contribute to the development of resistance. This presentation describes work using anionic microgels to create self-defensive surfaces. These charged microgels can be electrostatically deposited onto surfaces and subsequently loaded with small-molecule cationic antimicrobials by complexation processes. We have identified conditions where complexed antimicrobials remain sequestered for several weeks despite the fact that a diffusion-based elution process would take only seconds. These sequestered antimicrobials can, however, be released when closely approached by bacteria. Contact creates a steep chemical-potential gradient that drives antimicrobial transfer from the microgels to the bacterial envelope where there is a high concentration of negative charge. This transfer kills the bacteria. Metabolizing tissue cells in contact with the same microgel-modified surfaces are, however, unable to trigger antimicrobial release, indicating that the contact-transfer mechanism is specific to bacteria. Such specificity is critical for designing future biomaterials that not only promote healing but simultaneously resist infection.

531086 Investigating the Driving Forces of Assembly in Concentrated Electrolyte Solutions

Abstract:

The self-assembly of proteins, polymers, and nanoparticles is a complex phenomenon driven by multi-scale, hierarchical interactions with multiple association pathways. All-atom molecular simulations of these systems provide key, molecular-level insights into these processes. Typically, the high computational cost of simulating "realistic" systems is avoided by modeling small assemblies in dilute environments. However, these results provide limited insight and are often hard to corroborate with experiments.

In this talk, I will highlight how recently introduced advanced sampling techniques (metadynamics) can make simulations of crowded, self-assembling systems computationally tractable.[1,2] Using this sampling method, we are able to detect of stable clusters of 7 and 13 particle systems with faster convergence than previous metadynamics-based methods. Then, we apply this method to study electrolytes (NaCl and KCl) at different concentrations in solutions and near a surface (ideal mica surface). We find that the free energy of association of electrolytes with other electrolyte atoms, water, and the surface is sensitive to concentration. Since most theories of colloidal association only account for dilute concentration limits, we explore how these microscopic details of association may affect macroscopic association forces.[3]

REFERENCES

(1) Pfaendtner, J.; Bonomi, M. Efficient Sampling of High-Dimensional Free-Energy Landscapes with Parallel Bias Metadynamics. *J. Chem. Theory Comput.* **2015**, *11* (11), 5062–5067.

(2) Prakash, A.; Fu, C. D.; Bonomi, M.; Pfaendtner, J. Biasing Smarter, Not Harder, By Partitioning Collective Variables Into Families. (*Under Review*). **2018**.

(3) Prakash, A.; Pfaendtner, J.; Chun, J.; Mundy, C. J. Quantifying the Molecular-Scale Aqueous Response to the Mica Surface. *J. Phys. Chem. C* **2017**, *121* (34), 18496–18504.

528914 Chiromagnetic Nanoparticles and Gels

Abstract:

Chiral inorganic nanostructures have high circular dichroism, but real-time control of their optical activity has so far been achieved only by irreversible chemical changes. Field modulation is a far more desirable path to chiroptical devices. We hypothesized that magnetic field modulation can be attained for chiral nanostructures with large contributions of the magnetic transition dipole moments to polarization rotation. We found that dispersions and gels of paramagnetic Co_3O_4 nanoparticles with chiral distortions of the crystal lattices exhibited chiroptical activity in the visible range that was 10 times as strong as that of nonparamagnetic nanoparticles of comparable size. Transparency of the nanoparticle gels to circularly polarized light beams in the ultraviolet range was reversibly modulated by magnetic fields. These phenomena were also observed for other nanoscale metal oxides with lattice distortions from imprinted amino acids and other chiral ligands. The large family of chiral ceramic nanostructures and gels can be pivotal for new technologies and knowledge at the nexus of chirality and magnetism.

537730 Effects of Matrix Chain Length on Miscibility of Nanoparticles

Abstract:

Recent work in nanocomposites have studied ways to control the dispersion of nanofillers in polymer matrices/melts. The grafting of chemically similar polymer chains to the nanofiller has been a commonly proposed solution. However, aggregation of filler particles remains a problem. Furthermore, the particle aggregation observed experimentally by decreasing the graft/matrix chain ratio (N/M) remains unobservable in simulations of a single nanoparticle (NP) in a homopolymer matrix. Here, molecular dynamics simulations were used to study the miscibility effects of decreasing the ratio N/M on a polymer grafted nanoparticle in a polymer melt. The polymer matrix chain length was varied systematically in a series of simulations , while NP radius, grafting density, and grafted chain length were held constant. We first show that composition profiles of graft/matrix chains do not reveal any signs of brush collapse, observed experimentally, with increasing matrix chain length. The brush heights were then calculated using a second moment of the segment density and only highlighted a good to theta solvent transition. As most of the differences in segment density are evident closer to the surface of the nanoparticle, we thus conjecture that the first solvation shell, defined as the region bounding the first peak of the radial distribution function of monomers around the NP, might be more illuminating. Indeed, an analysis of the monomer fluctuations in the first solvation shell reveals several findings: (1) Miscibility effects by the matrix on the polymer grafted NP can be observed in the first solvation shell, (2) The matrix chains rather than the graft chains might give more insight into any conformational changes in the brush, (3) there is a subtle hint of a second transition at a N/M ratio of âˆ¼ 0.3 . These results imply that the interactions at the surface, perhaps between the NP itself and the melt, are significant in studying changes in miscibility of the polymer grafted NP and might be driving the morphology of the entire brush.

536107 Identifying Thermally and Kinetically Favorable Conditions for DNA-Mediated Assembly of Crystal Structures

Abstract:

Colloidal particle self-assembly using DNA functionalized particles (DFP) has been demonstrated as a promising way to assemble nano and micro particles for various applications,like catalysis, drug delivery, and molecular sensing. Grafting DNA molecules around nanoparticles allows for the precise control of inter and intra particle interactions, and this high degree of tunability is critical to program target structures. In this research, we used molecular dynamics simulations to investigate several design parameters which affects the ability of DFPs to form crystals. Using a coarse-grained DFP model with explicit nucleotide representation that incorporates hydrogen bonding and stacking interactions, we showed that design parameters such as the DNA chain length, temperature, and pairwise interaction strength affect the ability of DFPs to form crystal structures. Those design parameters were used to create order diagrams and identified the region in the design-space where crystal formation is favored. Additionally, we studied the kinetic properties of DFPs by tracking bond survival time and diffusivity. We found that longer DNA chains slow down overall kinetics and hinder the ability of DFPs to form crystal structures. The identified region where formation of crystals is kinetically-favored is in agreement with experimental results.

531076 Origins of High-Pressure Structural Stability, Elasticity and Self-Healing Property in Ligand Capped Nanoparticles Supercrystals

Abstract:

Ligand capped nanoparticles spontaneously assemble into supercrystals that have remarkable tunable properties. In these solids, the ligands, which link the metallic and semiconducting nanoparticles, play the central role in determining structural symmetry and their properties. These linker molecules can combine specific optical, electronic, or magnetic functionality of nanoparticles with the flexibility of self-assembly, and provide promising pathways for the development of tunable materials. However, the current development of these ligand decorated supercrystals is limited by many factors. First, a generic strategy to control the morphology, long-range order, lattice spacing and tailoring their properties is not yet established. Second, these materials exhibit a wide variation in its response to external conditions, particular pressure and deformation. The underlying microscopic origins of their unusual variation in mechanical properties remains unresolved. Furthermore, how and when failure of supercrystals under stress occur is an open question in materials science. Here, we employ large-scale molecular dynamic (MD) simulations to address these issues. We systematically investigate the role of ligands on the crystalline symmetry, pressure behavior, elasticity, deformation and self-healing properties of a model supercrystal, made of oleic acid capped PbS nanoparticles. The interplay between deflection and interdigitation of ligands are shown to be critical for crystalline symmetry and maintaining high-pressure structural stability. The elasticity viz., the Youngâ€™s modulus and bulk modulus are shown to be strongly correlated with the ligand concentration. Further, our MD simulations reveal molecular mechanisms of elastic deformation and healing in a supercrystal. We show inter-relationships between elastic deformation, self-healing ability and strain rate. These interconnections provide guidance towards the design of supercrystals with improved stability, and tunable mechanical property for future nanotechnology applications.

539214 Controlled Self-Assembly of Cationic Polyelectrolytes and Anionic Surfactants in Microfluidics Channels

Abstract:

The Nobel Prize-winning American theoretical physicist, *Richard Feynman*, pondered the potential of miniaturization in science and technology in his celebrated lecture at the 1959 meeting of the American Physical Society. His famous phrase "There is plenty of room at the bottom" beaconed the development of microelectronics and nanotechnology in the next 50 years. Computer chips with millions of conducting 'channels' for electrons are the cornerstone of our civilization today. It is fascinating that the same approach is now used in chemistry: polymer chips with microchannels inside can handle dozens of liquid flows simultaneously thus heralding the era of new chemistry for processing various reactants and nanoparticles. Similar to 'microelectronics' in physics, this approach is called 'microfluidics'.

Microfluidics is an interdisciplinary area combining chemistry, biology, fluid mechanics, nanotechnology and engineering. It deals with liquid flows in channels with the diameter between several micrometers and a millimeter, reduces dramatically the amount of required reagents; only minutes are needed to carry out complex reactions and analytic procedures which usually require weeks for preparation. In addition, microfluidics provides unique conditions for synthesizing and fine-tuning nanoparticles, as well as making drug delivery complexes which cannot be obtained in bulk solutions.

Polymers and surfactants are indispensable components of living matter and chemical industry processes: nucleic acids and proteins, detergents, enhanced oil recovery reagents, and drug delivery systems. They can easily form nanoparticles by self-organization. They are 'smart systems' as they can be easily controlled by many factors, such as pH, temperature, solvent, and etc. Thus, it is not a surprise that they are common objects of soft matter nanotechnology.

It is thrilling that practical application of such systems often involves <u>microchannels</u> (!): blood capillaries, micropores of catalyst carriers, 3D-printing systems, and etc. If we want to know how to apply them properly, we have to study them in microchannels. Microfluidics, however, mostly deals with flow hydrodynamics, fabrication of chips and chemical analysis. Polymer-surfactant systems are also mostly studied in bulk solutions, and not in microchannels. Several publications discussing polymers and surfactants in microfluidic devices are rather individual papers than a focused approach. Thus, a systematic and focused study of polymer-surfactant association in microfluidic environment will be an excitingly new interdisciplinary area which will contribute both to fundamental knowledge and practical applications of such systems.

This research is dedicated to synthesis and characterization of polyelectrolyte-surfactant nanocomplexes and to tuning their properties with the use of microfluidic approach.

The research is the joint project of Carnegir Mellon University and Kazan National Research Technological University (Russia).

This research project implementation was focused on the following objectives:

1. Designing and fabricating microfluidics devices for polymer-surfactant associations;
2. Synthesizing polymer-surfactant associates in microfluidic confinement through self-organization; characterization of complexes in bulk solutions;
3. Tuning of size and surface properties of polymer-surfactant complexes in microfluidic chips.

Research Methodology:

Microfluidic devices: PDMS-based chips made by microlithography method.

Polymers to be used for the research: chitosan and carboxymethyl cellulose (natural-occurring and eco-friendly polyelectrolytes of different charges).

Surfactants: sodium dodecyl sulfate and cetyltrimethylammonium bromide (model synthetic surfactants of different charges).

Project results:

Microfluidic chips for synthesis of polymer-surfactant nanocomplexes have been designed. The size of microchannels was selected in the 50-200 μm range.

Microfluidic chips were prepared using soft lithography method: PDMS samples with microchannels were bonded to glass plates using plasma processing to make the contacting surfaces hydrophylic.

Research methods: Microtensiometry; Optical Microscopy; Microviscometry; Soft Lithography (for Fabrication of microfluidics chips), Dynamic light scattering.

Individual fows of polymers and surfactants were studied in microchannels. The size of polymer nanoparticles was controlled by the solvent composition: alcohol solutions of various concentrations were added as co-solvents. Dynamic light scattering studies demonstrated that the size of polymer nanoparticles changed in the range of 50 - 200 nm, while microfluidic confimenement provided excellent environment to control the flows of polymer solutions.

Polymer-surfactant complexes have been synthesized in microfluidic channels. The following methods were used to tune the properties of these nanocomplexes:

- polymer-to-surfactant ratio provided control of sizes in the 20-100 nm range

- change of solvent composition by ading low-molecular salt and alcohol (20-200 nm range of nanoparticles)

Phase behavior of polymer-surfactant complexes was also studied. The study revealed that soluble and insoluble complexes form in microfluidic channels and formation of phases can be teracked by optical microscopy.

Polymer-surfactant nanocomplexes synthesized in microchannels demonstrate similar associative and phase behavior as the same complexes synthesized by conventional "in vitro" mixing. Microfluidics, however, is an excellent tool to accelerate synthesis of polyelectrolyte-surfactant nanoparticles and apply various automated methods to control their properties.

529320 Chemical Engineering Science for Graphene Technology Development

Abstract:

Early graphene research was led by condensed matter physicists focused on the electronic structure and properties of this newly isolated, single-atom-thick, two-dimensional (2D) material. Over the last decade graphene research has expanded greatly to include related sheet-like carbon forms and a wider range of properties and potential applications. The further development of large-scale graphene-based technologies will require contributions from a wide range of disciplines, including chemical engineering.

Many scalable graphene-based technologies use aqueous processing of graphene oxide nanosheets for assembling new carbon materials that include fibers, foams, and membranes. This approach of â€œbuilding with grapheneâ€ involves colloidal interactions and conformational changes of GO nanosheets, which are governed by weak intermolecular forces and result in mechanically soft and flexible final products.

Chemical engineering can play a key enabling role in the development of these technologies through application of the principles of colloid science, complex fluids, reaction engineering, and porous material transport. This talk will describe recent work on graphene oxide as (i) a high-aspect-ratio sheet-like colloid, (ii) an energetic material, (iii) a lyotropic discotic liquid crystal, (iv) a printable ink, and (v) a giant molecular precursor for liquid-phase assembly into engineered 3D carbons. The talk will also discuss applications of these novel carbons as flexible fibers, stretchable, breathable molecular barriers, selective transport membranes, and 2D templates for ceramic growth.

536011 From Energy Harvesting to Living Plants - Concepts in Biosensing and Energy Conversion Using Carbon Nanomaterials

Abstract:

Our lab at MIT has been interested in how the 1D and 2D electronic structures of carbon nanotubes and graphene respectively can be utilized to advance new concepts in molecular detection, as well as energy generation. By taking advantage of the exceptional electronic properties of these nano-structures, we continue to discover potential application spaces where carbon can play an important role. For example, we have pioneered a novel technique called Corona Phase Molecular Recognition, or CoPhMoRe, for discovering synthetic, heteropolymer corona phases that form molecular recognition sites at the nanoparticle interface. By screening libraries of synthetic heteropolymers chemically adsorbed onto single-walled carbon nanotubes (SWNT), we have engineered new optical biosensors that exhibit high selective recognition for bio molecules, such as riboflavin, L-thyroxine, dopamine, nitric oxide, sugar alcohols, estradiol, and fibrinogen. These results have significant potential for using SWNT-based sensors to interface to biological systems, allowing monitoring pathways at the sub-cellular, cellular, tissue, and whole-animal scale. I will also highlight our recent efforts in initiating an endeavor we call "Plant Nanobionics". There we use techniques to deliver and transport functional nanoparticles into living plants to grant them non-native functions. Our goal is to engineer plants to take over many of the functions now performed by electrical devices. I will introduce the nanoparticle co-localization mechanism in a plant, and highlight some of our recent nanobionic plant prototypes including a light-emitting plant. Lastly, I will briefly describe several applications of carbon nanomaterials in the energy space that have come out of our lab. There is a pressing need to find alternatives to conventional energy generation techniques, specifically those that rely on elements in finite global supply. We introduce Asymmetric Doping Cells (ADC), which convert chemical potential to electrical energy by means of spatially selective doping along a nanostructured conduit or particle. These ADCs have applications to energy harvesting from aqueous and organic solvents, as well as electro-catalysis for chemical synthesis. An inverse length-scaling of the maximum power as $L^{-1.03}$ that creates specific powers as large as 30.0 kW kg^{-1} highlights the potential for microscale energy generation. We also introduce carbon materials for what we call thermal resonators that make use of thermal storage elements with high effusivity – the product of the thermal conductivity and heat capacity to the one half power. Thermal resonators base on carbon can harvest energy indefinitely from ambient thermal fluctuations of various frequencies, opening new possibilities for remote applications.

537883 Demineralized Bone Slices for in Vitro Endosteal Niche Modeling

Abstract:

The trabecular bone implicates clinical significance as a major site for preserving adult stem cells, generating all types of blood cells throughout lifetime, osteoporosis during aging, and in case of cancerous lesion metastasis. In vitro reconstitution of trabecular bone tissue models is imperative to better understand fundamental biology of aforementioned biological processes and explore translational opportunities such as ex vivo expansion of hematopoietic stem cells and drug screening for cancer metastasis and osteoporosis. In this tissue, bone and marrow exist as an integrated organoid. Each tissue represents distinct cellular and extracellular niches that are known as vascular and endosteal niches. While there has been significant progress in modeling vascular niche, in vitro reconstitution of endosteal niche remains premature. The field has focused on bone-forming osteoblasts and bone-resorbing osteoclasts in in vitro bone tissue modeling. However, in reality, the major cell types comprising the endosteal niche is osteocytes and bone lining cells. We have developed a new biomaterial and engineering strategy to faithfully create endosteal bone tissue models. Thin slices of demineralized bone matrix supported osteoblast cell activity and their lining cell differentiation - inactive form of osteoblasts. Further stacking demineralized bone matrix allowed creating lamellar structure of bone, which in turn promoted osteocyte differentiation of osteoblast. We provide evidence that support bone lining and osteocyte differentiation with related gene expression, cellular activities and morphology and secretory profiles. The established endosteal niche model is expected to serve as a foundation for creating better functional trabecular bone tissue models combined with existing vascular niche models.

534725 Defining the Mechanisms of Immune Resolution after Biomaterial Implant into Adipose Tissue

Abstract:

Drug delivery to the adipose tissue is a promising therapeutic strategy for metabolic diseases associated with this tissueâ€™s dysfunction, such as obesity and type two diabetes. Current pharmacological approaches to treat metabolic disease involve oral administration of drugs that are rapidly cleared in the liver or broken down in the gut, which cause side effects that negatively impact patient compliance. To circumvent these issues, we propose to utilize biodegradable polymer scaffolds for drug delivery to the adipose tissue. This strategy allows for high drug concentrations in the fat and protects the patient from side effects. Once implanted, the scaffold will interact with resident immune cells, which play a central role in the regulation of metabolism and are dysfunctional during obesity. There is limited knowledge regarding how the adipose immune environment responds to biomaterial implants and how this response can be harnessed to alter adipose tissue function. To address these knowledge gaps, porous poly(lactide-co-glycolide) (PLG) scaffolds, fabricated from microparticles using gas foaming, were implanted into the epididymal fat pad of C57BL/6 mice and the immune response was characterized at various time points. Flow cytometry indicated an influx of macrophages, neutrophils, monocytes, and lymphocytes 7 days after implant. Interestingly, immune cell numbers dramatically decreased between 7 and 14 days after implant, with immune cell populations returning to pre-implant levels. This presentation will focus on the mechanism by which the immune response to scaffold implant resolves in adipose tissue. We will discuss expression of cytokines, adipokines, and apoptosis proteins at the implant site over the time frame of the immune response. Furthermore, we will present results of experiments designed to understand if this response is unique to PLG and fat by utilizing polycaprolactone scaffolds and investigating subcutaneous implant sites. Collectively, this work will provide an understanding of how the adipose immune environment responds to a PLG-based drug delivery device and lays the groundwork for future studies to direct this immune response to modulate adipose tissue function as a treatment for metabolic disease.

538066 A Poly-L-Lactide Scaffold with Continuous Gradient Pore Size That Differentially Induce Local Chondrogenesis and Osteogenesis for Osteochondral Repair

Abstract:

INTRODUCTION: Cartilage and bone in articular joints are intimately linked, and a better understanding of disease and regenerative processes calls for the simultaneous investigation of both tissues as part of the osteochondral (OC) unit. Furthermore, scaffold-based regenerative approaches in the joint often target both cartilage and the subchondral bone, taking advantage of the endogenous bone marrow stem cells made available by breaching through the OC junction. However, the production of scaffolds for OC regeneration is challenging, as scaffolds must provide mechanical strength while also mimicking the local cartilage and bone microenvironments. Thermally Induced Phase Separation (TIPS) is one of the most adaptable techniques to produce porous scaffold for tissue engineering (1). A wide range of morphologies in terms of pore size and distribution can be obtained by tuning TIPS processing parameters, primarily thermal history (2). In this work, TIPS was used to produce a poly-L-lactide (PLLA) scaffold for OC tissue engineering with a continuous pore size gradient along the sample thickness, starting at ~70 μm diameter on one side and increased steadily to ~200 μm diameter on the opposite surface. We expect that the smaller pore size will support the induction of chondrogenesis while the larger pore size will induce a more osteogenic phenotype.

A microphysiological tissue system (MPS) bioreactor has been developed to replicate *in vitro* the *in vivo* OC physiological conditions (3). The MPS allows separate control of the chondral and osseous environment while permitting communication between chondrocytes and osteoblasts across the OC junction (4), similar to the conditions of OC tissue *in vivo*. We have used here our MPS system to validate the TIPS-generated pore-gradient PLLA scaffold.

MATERIALS AND METHODS: PLLA (Resomer TN L 209 S), 1,4-dioxane (Sigma-Aldrich) and distilled water were used to produce the polymeric foams. The ternary solution employed has a PLLA concentration of 4 % wt, in 87:13 wt/wt dioxane/water (solvent/non-solvent). Different thermal histories were imposed on sample sides by an experimental apparatus made in-house using software controlled Peltier cells. Starting from a homogeneous system, i.e. well above the cloud point temperature (41 °C), each sample surface was brought to the same demixing temperature (30 °C) at different cooling rates: 25°C/min (Fast Cooling Side – FCS) and 1°C/min (Slow Cooling Side – SCS). The residence time inside the metastable region for each side was set at 50 and 20 min, respectively, creating a monotonously variable pore size along sample thickness. Then, both surfaces were rapidly cooled to -20°C to freeze the as-obtained microstructure, and the solvents were extracted by rinsing in distilled water and drying under vacuum. Cylindrical scaffolds were punched out and uniformly seeded with 800,000 bone marrow hMSCs derived from total joint arthroplasty patients (IRB: University of Washington), and then placed in the MPS with the small pore side (~70 μm) in the upper chamber (UC) and the large pore size in the lower chamber (LC). After 7 days of growth medium (GM: DMEM, 10%fetal bovine serum (FBS), penicillin/streptomycin/fungizone) perfusion at 2 ml/day, chondrogenic medium (CM: DMEM, 10 ng/ml TGF-β3 (PeproTech), 1% insulin-transferrin-selenium, 50 μM L-ascorbic acid 2-phosphate (AsA2-P), 40 μg/ml L-proline) was supplied to the UC, and osteogenic medium (OM, GM + 0.1 μM dexamethasone, 50 μg/ml AsA2-P, 10 mM β -glycerophosphate) to the LC, both at a flow rate of 2 μL/s. After 4 weeks of differentiation, engineered OC tissues were collected and analyzed by RT-PCR and histology.

RESULTS: At 4 weeks, the constructs exhibited strong alcian blue staining on the side residing in the UC (exposed to CM) and alizarin red staining in the part residing in the LC (exposed to OM). Constructs cultured only in CM (same medium in both chambers) exhibited stronger alcian blue staining on the small pore side, while correspondingly stronger alizarin red staining was observed in constructs uniformly cultured in OM. OC constructs fabricated in photocrosslinkable methacrylated gelatin using the same hMSCs and cultured in the same conditions within the MPS were used as 3D control. Histological staining was significantly stronger for the PLLA OC constructs. RT-PCR analysis of PLLA constructs cultured with CM in the UC and OM in the LC confirmed upregulation of chondral genes (COL2, ACAN, SOX9) in the upper part of the construct (small pores) and of osseous genes (RUNX2, BSPII, OPN) in the lower part of the construct (large pores), compared to day 0. Similar gene expression profiles for the parts in the UC (CM) and LC (OM) in the PLLA and methacrylated gelatin OC constructs were found.

DISCUSSION: A spatially defined, biphasic differentiation of hMSCs within the engineered PLLA OC constructs is observed. Small pore size in the scaffold favored chondrogenic differentiation, likely because of enhanced cell-cell contact inducing mesenchymal condensation, while larger pore size favored osteogenic differentiation, possibly due to easier nutrient movement and higher biosimilarity to the osseous, trabecular microenvironment. The PLLA pore-gradient structure is thus a promising scaffold for OC repair. Furthermore, even without application of osteogenic factors, hMSCs seeded in the LC (OM) of the scaffold presented good osteogenic differentiation, confirming the validity of the MPS design.

Overall, a single unit biodegradable scaffold produced with local gradient pore structure may be tailored for chondro- and osteoinduction and applied for OC repair to improve osteochondral tissue repair, providing mechanical stability and local cues for cell differentiation.

REFERENCES: (1) Mannella GA. *J Polymer Sci B: Polymer Physic*. 2014, 52(14): 979–983. (2) Mannella GA et al. *Mater Letts*. 2015, 160: 31-33. (3) Lozito TP et al. *Stem Cell Res Ther* 2013, 4(Suppl 1):S6. (4) Lin H et al. *Mol Pharmaceut*. 2014, 11(7): 2203-12.

ACKNOWLEDGEMENTS: Commonwealth of Pennsylvania, NIH (1UG3 TR002136-01), Ri.MED Foundation.

Fig.1 (A) Experimental design & MPS. (B) Histology: PLLA and gelatin constructs in MPS (dual media). (C) PLLA cultured in only CM (pore size effect).

538434 BMP-2 Conjugated Micro-Fiber/Hydrogel Composites for Bone Integration to Engineered Ligament Tissue

Abstract:

While the bone-patellar tendon-bone autograft is considered the gold standard material for anterior cruciate ligament (ACL) reconstruction, alternative materials such as the hamstring or quadriceps tendon reveal that directly fixing the soft tissue graft to the native bone via bone screws risks mechanical instability and pullout of the graft. This is intrinsically a materials problem whereby micromotions and stress concentrations between the hard bone tissue and soft implant impede bone integration. In view of this challenge, the purpose of this work is to develop an engineered bone regeneration structure – as a portion of an engineered ligament tissue – that could be easily infiltrated by new bone tissue to facilitate integration of the implant within the bone tunnels.

To this end, we applied a two-step approach. First, we immobilized bone morphogenic protein (BMP)-2 to the surface of electrospun microfibers and characterized its ability to stimulate osteoblastic differentiation of bone marrow-derived mesenchymal stem cells (BMSCs). Specifically, we examined the effect of incorporating 0.5 and 1% wt% heparin into polycaprolactone (PCL) fibers on BMP-2 adsorption and covalent conjugation (using EDC/NHS bioconjugation). Here, preliminary data show a five-fold increase of immobilized BMP-2 concentration at the highest heparin content when covalently tethered to the fibers. In addition, *in vitro* culture of BMSCs demonstrated that the micro-fiber meshes support cell viability and proliferation. In addition, alkaline phosphatase activity – a marker of osteoblastic differentiation –was significantly higher upon the binding of BMP-2 to the meshes. PCR assays to assess other osteogenic markers are underway.

Second, to form a robust 3D framework for bone-regeneration, we embedded this osteogenic fiber mesh into a cross linked 3 % gelatin methacrylate (GelMA) hydrogel. Here, a simple multilayered fiber/hydrogel composite system was created and the migration of MSCs through the hydrogel and onto the fiber layer in response to BMP-2 was assessed by confocal imaging. In the same system, the expression of markers of osteoblastic differentiation will be analyzed by real-time PCR. We postulate that the porosity provided by the GelMA phase, coupled with the mechanical properties and localized bioactive signals of the BMP-2 conjugated fibers will facilitate bone formation both *in vitro* and ultimately *in vivo*.

535576 Human Skeletal Muscle Growth and Maturation in 3-Dimensional Silk-Extracellular Matrix Scaffolds

Abstract:

Injuries resulting in volumetric skeletal muscle loss (VML) are unable to heal and properly regenerate, resulting in a permanent functional deficit. Advances in tissue engineering have created new opportunities for treatments that use engineered muscle to replace the lost tissue. However, a thorough understanding of the material properties that lead to healthy skeletal muscle tissue regeneration as opposed to scar tissue formation is needed. The process of fibrosis and scar tissue formation following muscle injury relies on many variables, some of which are dependent upon the patientâ €™s own wound healing rates and some which are dependent upon the formulation of the biomaterial implanted. Furthermore, previous work in our lab has demonstrated that the addition of bioactive components, such as extracellular matrix (ECM) and growth factors, to silk scaffolds alters *in vivo* degradation and resorption. Therefore, we aimed to understand the role of biomaterial formulation on muscle, vascular, and stromal cell behavior *in vitro* to guide future *in vivo* applications for repair of skeletal muscle defects. Specifically, we developed three-dimensional silk fibroin-based tissue constructs containing skeletal muscle derived ECM and growth factors (vascular endothelial growth factor VEGF, epidermal growth factor EGF). Engineered muscle tissue was formed using human skeletal muscle cells with and without human endothelial cells (HUVECs, Lonza®) to create engineered tissues with desired properties for future *in vivo* applications. Cell phenotypes within the scaffold were evaluated by immunohistochemistry and Western blot. The results showed that the method of growth factor presentation during 3D culture impacted human skeletal myoblast phenotype and maturation through shifts in desmin, myogenin, pax7, and alpha-actinin expression in human skeletal muscle and CD-31 (PECAM-1), von Willebrand factor, and VE-cadherin expression in human umbilical vein endothelial cells. Therefore, the rational design of 3D scaffolds for implantation *in vivo* based on *in vitro* analyses should improve engineered tissue integration, and the patientâ€™s muscle function following biomaterial resorption and degradation. Preliminary results confirmed biomaterial formulation impacts engineered primary rodent skeletal muscle pre-conditioning within a bioreactor system, resulting *in vivo* functionality following pre-conditioned scaffold implantation.

529118 3D Graphene Foam Based Scaffolds to Control Transdifferentiation of MSCs into Schwann Cell-like Phenotypes Via Electrical Stimuli for Peripheral Nerve Regeneration

Abstract:

Multipotent mesenchymal stem cells (MSCs), isolated and derived from various connective tissue sources and transdifferentiated into Schwann cells (SCs), hold considerable potential for cell-based peripheral nerve regeneration therapies using autologous transplantation via nerve guidance conduits (NGCs).[1] However, current problems regarding; 1) the efficiency of conventionally applied reversible and unscalable, chemical stimuli-based transdifferentiation protocols; 2) controlling final fate of the implanted cell population; and 3) design of 3D microstructure of scaffolds completely mimicking the extracellular matrix, limit the clinical use of transdifferentiated MSCs.[1] Therefore, there is a critical need for alternative approaches to gain mechanistic understanding of transdifferentiation in 3D scaffold microenvironment that can enable precise control on the final fate of the implanted cell population, enhanced paracrine activity, myelination, axonal regeneration and functional clinical outcomes.[1]

In previous studies, we have successfully demonstrated transdifferentiation of MSCs to SCs fates displaying a high degree of SC-markers immunolabeling and neurotrophic factor paracrine activity.[2-5] We showed MSCs to SCs transdifferentiation via a chemical induction paradigm on 2D Poly-L-Lactic Acid (PLLA) films.[3, 6] In addition, we also demonstrated that PLLA-based 2D platforms with surface micropatterns and nerve growth factor (NGF) gradients can be used as depots for the controlled release of NGF, mimicking NGF secretion at physiological levels and promoting directed neurite extension from PC12 cells.[6] Considering that 3D platforms, mimicking the natural ECM microenvironment, would help to obtain more realistic information, we fabricated gelatin conduits possessing different 3D microstructures and mechanical properties to control the direct transdifferentiation of MSCs into SCs within the 3D conduit matrix via chemical induction.[4] Recently, using comparative proteomics, we identified significantly regulated proteins and related cellular pathways upon transdifferentiation of MSCs into SCs via chemical stimuli on 2D substrates.[5] To eliminate the need for expensive chemicals/growth factors and provide a more precise platform to control MSCs transdifferentiation and fate commitment, we demonstrated the first proof-of-concept for the MSCs to SCs transdifferentiation via sole electrical stimulation (100 mV, 50 Hz, 10 min per day during 15 days), free from any chemical stimuli, using conductive graphene-based 2D substrates.

Based on this background, in this study, an alternative strategy based on PLLA films with 3D porous microstructure, surface micropatterns and integrated graphene micro-circuits is used in combination with 3D conductive graphene foams possessing desirable 3D microstructural/mechanical properties to provide an available cellular microenvironment and electrical transdifferentiation. Briefly, PLLA films with micropatterns were prepared as mentioned previously[6] and graphene-based integrated circuit designs are transferred to the film surface. Then, 3D graphene foam, prepared using chemical vapor deposition (CVD) is integrated into graphene circuits on the film surface via in situ adhesion. The films can easily be rolled into conduits in such a way that the surface with micro-circuits and graphene foam becomes the inner lumen and may be used for *in vivo* applications. The graphene foams and PLLA films in combination provides a desirable 3D microstructure and a permissive microenvironment for cells and enables precise control in electrical stimuli through the integrated micro-circuits and conductive foams for cellular transdifferentiation. Our results indicated that MSCs seeded on graphene foams showed good attachment, proliferation and growth within the 3D foam microstructure for 7 days with high cell viability (~97% cell survival â€" based on Trypan blue staining). The seeded MSCs were transdifferentiated into SCs via chemical[4] or electrical stimuli[2] as described previously. The immunocytochemical (ICC) studies indicated that ~80% of the cells showed successful transdifferentiation (labeling for SC markers s100, s100Î² and p75) for both chemical and electrical stimuli compared to the non-transdifferentiated control MSCs. The transdifferentiated MSCs secreted significantly higher amount of growth factors as determined by ELISAs (particularly NGF: ~ 2.5 pg/mL per cell) for both chemical and electrical stimuli based transdifferentiation approaches, which was significantly higher than the non-transdifferentiated control groups. These results were considerably better than the electrically transdifferentiated cells on graphene-based 2D substrates (~75% of the MSCs differentiated into SCs like phenotypes and ~0.55 pg/mL per cell of NGF secretion) as reported in our previous work[2] and comparable to the chemically transdifferentiated MSCs within 3D gelatin scaffolds (~85% of the MSCs differentiated into SCs like phenotypes and ~2.5 pg/mL per cell of NGF secretion) as indicated in our previous work.[4] The ICC and ELISA findings are further supported by RT-PCR and Western Blot (WB) experiments performed using selected primers and antibodies to detect the regulated genes and proteins. The co-culture of MSCs seeded in the proposed conduit design with PC12 cells is also evaluated to assess the functional biological activity of secreted growth factors. Overall, these results suggest that the electrical stimuli applied within the 3D graphene matrix enables comparable differentiation and paracrine activity with transdifferentiation procedures involving expensive chemical stimuli applied in 3D scaffolds[4] or 2D substrates[2], ultimately leading to development of novel and promising nerve regeneration strategies.

References:

1. Uz, M.; Das Suprem, R.; Ding, S.; Sakaguchi Donald, S.; Claussen Jonathan, C.; Mallapragada Surya, K., Advances in Controlling Differentiation of Adult Stem Cells for Peripheral Nerve Regeneration. *Advanced Healthcare Materials 0* (0), 1701046.
2. Das, S. R.; Uz, M.; Ding, S.; Lentner, M. T.; Hondred, J. A.; Cargill, A. A.; Sakaguchi, D. S.; Mallapragada, S.; Claussen, J. C., Electrical Differentiation of Mesenchymal Stem Cells into Schwann-Cell-Like Phenotypes Using Inkjet-Printed Graphene Circuits. *Advanced Healthcare Materials* **2017**, n/a-n/a.
3. Sharma, A. D.; Zbarska, S.; Petersen, E. M.; Marti, M. E.; Mallapragada, S. K.; Sakaguchi, D. S., Oriented growth and transdifferentiation of mesenchymal stem cells towards a Schwann cell fate on micropatterned substrates. *Journal of bioscience and bioengineering* **2016,** *121* (3), 325-35.
4. Uz, M.; Buyukoz, M.; Sharma, A. D.; Sakaguchi, D. S.; Altinkaya, S. A.; Mallapragada, S. K., Gelatin-Based 3D Conduits for Transdifferentiation of Mesenchymal Stem Cells into Schwann Cell-like Phenotypes. *Acta Biomaterialia.*
5. Sharma, A. D.; Wiederin, J.; Uz, M.; Ciborowski, P.; Mallapragada, S. K.; Gendelman, H. E.; Sakaguchi, D. S., Proteomic analysis of mesenchymal to Schwann cell transdifferentiation. *Journal of Proteomics* **2017,** *165*, 93-101.
6. Uz, M.; Sharma, A. D.; Adhikari, P.; Sakaguchi, D. S.; Mallapragada, S. K., Development of multifunctional films for peripheral nerve regeneration. *Acta Biomater* **2016**.

538585 Nanofibrous Scaffolds Produced By Electrospinning, Rotary-Jet Spinning and Airbrush for Orthopedic Tissue Regeneration

Abstract:

Despite the progress tissue engineering has made in the development of improved biomaterials, inhibiting bacterial infection has not been a central focus to date. Infection is a leading cause of implant failures with many agencies (such as the Centers for Disease Control) predicting more deaths from bacteria than all cancers combined by 2050.Gram-negative bacteria are naturally resistant to numerous treatments and are difficult to kill due to their robust and hydrophobic outer lipopolysaccharide membrane which helps to prevent the flow of antibiotics or drugs into the cell. Moreover, due to extensive antibiotic use, gram-positive *Staphylococcus aureus* has evolved to a methicillin-resistant strain, which can overcome other classes of antibiotic treatments. The development of an implant capable of reducing bacterial growth (without resorting to the use of antibiotics, which causes antibiotic resistant bacteria) would be an effective way to improve implant success. Recently, scientists have been investigating novel materials and techniques to meet growing orthopedic tissue engineering needs. Due to problems (such as infection and long healing time) that titanium based implants cause *in vivo*, new polymeric materials have been introduced. The first step in implant infection is bacterial adhesion, which can potentially result in the formation of antibiotic resistant biofilms for some species. Bacterial adhesion, growth, and subsequent biofilm formation on surfaces are particularly resistant towards the body's defense mechanisms and antibiotic treatments, which can cause implant rejection. Multiple substrate properties, including chemical composition, hydrophobicity, and surface roughness, are believed to be of significance in the bacterial attachment process. Recently, researchers have been exploring alternative physical methods through reduced contact of microorganisms with an implant which would prevent extensive bacteria proliferation on the surface and allow the immune system to clear such bacteria. Materials with biologically-inspired nanostructured features have shown promise to reduce bacteria growth. Polycaprolactone (PCL) is a bioresorbable and biocompatible polymer with potential applications for bone and cartilage repair. In this study, polycaprolactone fibers (with and without hydroxyapatite nanoparticles (nHAp) and carbon nanotubes (CNT)) were produced using three different methods: electrospinning, rotary-jet spinning and airbrush. The scaffolds were characterized using contact angles, differential scanning calorimetry, scanning electron microscopy, and were subjected to cell culture, bacterial assays, and mechanical (tensile) testing. Early interactions between the surface of the material with biological fluids play a fundamental role in determining cellular behavior (such as adhesion, differentiation, and tissue formation). The biological and material properties were studied to understand how the various fabrication techniques and nanoparticles affect human osteoblasts, gram-positive *(Staphylococcus Aureus)* and gram-negative *(Pseudomonas aeruginosa)* bacteria growth on samples. Experiments showed no toxic effect on osteoblast cells and a significant decrease in bacterial density by adding nHAp and CNT to the PCL scaffolds without using growth factors or antibiotics. Most importantly, results showed that without the use of antibiotics, by only changing the fabrication method and the size of the fibers, antimicrobial properties were improved without sacrificing mechanical properties and osteoblast functions. Among the different fabrication methods, airbrushed fibers indicated the highest elastic modulus compared to the other two techniques. Specifically, it was found that the antimicrobial properties were improved when using airbrushing compared to electrospun and rotary-jet spun scaffolds, where up to a 50% decrease in bacterial density in both gram-positive and gram-negative species was observed (**Fig. 1**). To assess osteoblast biocompatibility and differentiation, cell proliferation, viability, adhesion, and calcium deposition assays were conducted. Although the cell proliferation, calcium deposition and cell viability results are similar, it is observed that the cell adhesion in airbrushed samples are poorer than those in electrospun and rotary-jet spun scaffolds. To investigate the influence of fibers' morphology on osteoblast cells, further gene expression assay (qPCR) will be conducted; which means based on the cell culture assays that have already been done, we should be looking for specific gene expressions such as alkaline phosphatase, and osteocalcin and compare them to the results from mammalian cell culture from different fibers. In other words, the future dream of this study is to demonstrate how different fabrication methods and hence various surface properties can make cells behave differently. It is of significance to mention that PCL fibers cannot compete with metallic implants such as titanium and stainless steel strength-wise. The application of this study is to heal the defects in bone/cartilage by increasing the tissue growth and minimizing bacterial colonization without use of any growth factor or antibiotic.

../Screen%20Shot%202018-04-18%20at%205.39.02%20PM.png

Figure 1. All samples contain 1% of CNT and 1% of nHAp. Colony Forming Unit (CFU) method was used to measure the density of bacteria. a) *Staphylococcus aureus* density, and b) *Pseudomonas aeruginosa* density in different samples. Values are mean ± SD, N=3, * $p < 0.05$, ** $p < 0.1$, *** $p < 0.01$, **** $p < 0.001$ compared to each other.

533769 Tissue Origami for Biomineralization

Abstract:

< style="padding-left: 22pt; text-align: center;">TISSUE ORIGAMI FOR BIOMINERALIZATION

Gulden Camci-Unal

University of Massachusetts Lowell, Department of Chemical Engineering, 1 University Avenue, Lowell, MA 01854, USA

Due to disease, degeneration, trauma, or aging, loss or damage to bone occurs in the body. Although there have been significant improvements in development of bone scaffolds, it remains difficult to fabricate porous and biocompatible constructs in physiologically relevant sizes in cm-scale. Thus, there is an unmet need for development of inexpensive, easy-to-use, and widely accessible biomaterials that are capable of supporting biomineralization for bone repair. In this work, we developed biomineralized origami-inspired paper scaffolds in three-dimensions (3D). This work is the first demonstration that paper can be used as a 3D construct to induce template-guided mineralization by osteoblasts.

We used the principles of origami to fabricate free-standing paper scaffolds in cm-scale. Because paper is an extremely flexible material that can easily be cut, creased, and folded to form 3D constructs, the scaffolds were easily fabricated in a variety of different geometries. This feature can be highly useful in fabricating constructs for patient-specific applications especially for patient who have defects of irregular sizes and shapes. Whatman brand filter paper was employed due to its large pore size (25-30 micron) and thickness (190 micron) for cellular migration. After sterilizing the constructs, they were seeded with mouse osteoblasts at a density of 20 million cells/mL in a collagen matrix (2 mg/mL). The samples were cultured up to 21 days and mineralization was evaluated using various assays including colorimetric assays, immunocytochemistry, high-resolution imaging (SEM), and micro-computed tomography (micro-CT) at different time points (days 0, 3, 7, 14, 21). We also performed in vivo experiments in a rat model.

In this work, for the first time, we performed tissue origami using paper for template-guided mineralization. We generated paper scaffolds in different shapes, sizes, and configurations. Due to its porous structure, paper allowed for transport of oxygen and nutrients across its thickness. Paper scaffolds supported a uniform distribution of cells within their 3D structures. In our experiments, proliferation of osteoblasts increased until day 3 and then decreased. Because cells proliferate first and then differentiate, a decrease in proliferation is expected to correlate with the initialization of the mineralization process. Hydroxyapatite content of the samples indicated that there was a progressive increase in the amount of hydroxyapatite in the paper scaffolds over 21 day of culture period. We found that the expression of osteocalcin increased until day 14 and decreased. This result can be attributed to increasing mineralization and consequent transformation of osteoblasts to osteocytes after day 14. We used SEM to visualize the deposition of mineral clusters, and EDAX to calculate the ratio of calcium to phosphate, which was found to be 1.5. Our in vivo experiments demonstrated that paper substrate did not cause inflammation. The paper implant integrated with the existing tissue well and rapidly vascularized.

In summary, we have demonstrated origami-inspired tissue engineering for mineralized tissues. We successfully obtained partially mineralized scaffolds in various 3D geometries. Mouse osteoblasts deposited calcium phosphate in these scaffolds and induced template-guided mineralization. Our approach used paper, a readily available material, as the cell culture scaffold. Paper has great potential to tackle the limitations of traditional scaffolds including cost, availability, accessibility, porosity, flexibility, rigidity, and ease of fabrication. In the future, paper-based scaffolds could potentially guide and accelerate bone repair using patient specific cells.

528383 Invited Speaker: Exciton Engineering for Brain Nanosensor Delivery, and Imaging of Modulatory Neurotransmitters

Abstract:

Neurons communicate through chemical neurotransmitter signals that either terminate at the postsynaptic process ("wired transmission") or diffuse beyond the synaptic cleft to modulate the activity of larger neuronal networks ("volume transmission"). Molecules such as dopamine belong to the latter class of neurotransmitters, and have been the pharmacological targets of antidepressants and antipsychotics for decades. Owing to the central role of dopamine over a range of behaviors and psychiatric disorders, real-time imaging of the signal's spatial propagation would constitute a valuable advance in neurochemical imaging. To this end, we present the synthesis and delivery of a nanoscale near-infrared fluorescent nanosensor for dopamine and demonstrate its efficacy for imaging dopamine volume transmission in the extracellular space of the brain striatum and prefrontal cortex with concurrent stochastic simulation validation [1]. The sensor is developed from polymers pinned to the surface of single wall carbon nanotubes (SWNT) in which the surface-adsorbed polymer is the recognition moiety and the carbon nanotube the fluorescence transduction element. Excitonic transitions in functionalized SWNT yield up to $\Delta F/F = 4500\%$ near-infrared fluorescence emission in the presence of dopamine. We next demonstrate imaging of evoked dopamine release in acute striatal slices, and show disrupted dopamine reuptake kinetics when brain tissue is exposed to antidepressant dopamine reuptake blocker nomifensine. Lastly, we discuss a new form of imaging for deep-brain neuromodulator detection: double infrared excitation-emission fluorescence microscopy [2]. We characterize our findings in the context of their utility for high spatial and temporal neuromodulator imaging in the brain, describe nanosensor exciton behavior from a molecular dynamics (MD) perspective, and validate nanosensor for use ***in vivo*** to correlate external stimuli (experiences, behavior) to chemical output (neurotransmission).

References

1. Beyene, A.B., McFarlane, I.R., Pinals, R.L, Landry, M.P. Stochastic Simulation of Dopamine Neuromodulation for Implementation of Fluorescent Neurochemical Probes in the Striatal Extracellular Space. ***ACS Chemical Neuroscience*** (2017). DOI: 10.1021/acschemneuro.7b00193

2. Del Bonis O'Donnell, J.T., Page, R.H., Beyene, A.G., Tindall, E.G., McFarlane, I.R., Landry, M.P. Molecular Recognition of Dopamine with Dual Near Infrared Excitation-Emission Two-Photon Microscopy. ***Advanced Functional Materials*** (2017). DOI: 10.1002/adfm.201702112

524763 Biodistribution and Drug Release Kinetics of Gold Nanoconjugates for Respiratory Recovery after Spinal Cord Injury

Abstract:

This collaborative research developed a patented nanodevice for targeted drug delivery to induce respiratory recovery after spinal cord injury. The nanodevice consists of a gold nanoparticle drug carrier chemically conjugated to methylxanthine drugs, such as theophylline and 1, 3-dipropyl-8-clyclopentyl xanthine, and a transporter protein (wheat germ agglutinin coupled to horseradish peroxidase or WGA-HRP). When theophylline was tested at the clinical level, most spinal cord injured patients could not tolerate it side effects when it was delivered systemically at standard therapeutic dose levels. We have demonstrated a strategy of selectively targeting the lower motor (phrenic) and pre-motor rostral ventral respiratory group (rVRG) neurons responsible for diaphragmatic function in rats by showing persistent respiratory recovery 48 hours after a single diaphragmatic injection of the nanodevice at 0.1% free drug dosage. In order to translate this technology for clinical use, we carried out pharmacokinetic and tissue distribution studies (e.g., gold accumulation) in an animal model. The amount of gold was determined by inductively coupled plasma mass spectrometry in the blood, RES organs, and targeted tissues after injecting the nanoconjugate into the left hemidiaphragm of adult female and male Sprague Dawley rats. Our data show that the 4 nm gold conjugated to the drug and protein 1 week post injection accumulates at a significantly higher dosage in targeted tissues including the spinal cord and medulla in the brain than unconjugated gold. We also carried out drug release studies from the nanodevice in buffers and artificial cerebrospinal fluid (CSF) for time periods ranging from 1 hour to 4 weeks by LC-MS/MS. The data show that the drug is slowly released from the gold nanoparticle carrier with less than 10% released after two weeks. These data may explain the long term recovery in the in vivo animals up to 4 weeks after the injection of the nanoconjugate. The translational significance of the work is the recovery of the diaphragm in spinal cord injured rats induced by the nanodevice at a fraction of the systemic drug dosage with the promise to mitigate drug side effects.

534100 Liposomes Functionalized with Cell-Penetrating Peptides As a Novel Treatment for Bacterial Meningitis

Abstract:

Liposomes Functionalized with Cell-Penetrating Peptides as a Novel Treatment for Bacterial Meningitis

Caterina Bartomeu, Di Shi, and Thomas J. Webster

Department of Chemical Engineering, Northeastern University, Boston, MA, USA

Bacterial meningitis is a very serious infection that can lead to permanent disabilities and death in as little as a few hours if left untreated. It is a brain inflammatory disease caused by inflammation of the meninges and in-between fluid-filled spaces. Severe cases can lead to permanent brain damage as well as death. Brain inflammatory diseases have become a global concern in clinical care due to the recent emergence of antibiotic-resistant bacteria, making it increasingly difficult to treat these infections. By forming a complex polysaccharide matrix known as biofilms, bacteria protect themselves from environmental and external stimuli such as antibiotics, allowing them to spread their antibiotic resistance within the bacterial community. The current broad-spectrum antibiotics face great difficulties in penetrating these biofilms, leading to ineffective treatment and progressed infection. In addition, the overuse of antibiotics promotes resistance in bacteria and leads to longer stays in hospitals and poorer patient care outcomes. Therefore, there is a need to develop alternative treatments for brain infection diseases such as bacterial meningitis.

To address this need, we report here the use of pH-sensitive liposomes as drug carriers to deliver targeted antibiotics at the site of infection. The ability of pH-sensitive liposomes to fuse with the bacterial cell membrane has been reported previously,[1] as well as the successfully encapsulation of different particles, such as antibiotics and metal nanoparticles, inside their vesicles or in-between their phospholipid bilayers.[2] Liposomes were functionalized with a cell-penetrating peptide, TAT (47-57) in order to be able to cross the blood-brain barrier (BBB), which is one of the main challenges for treating brain diseases.[3]

Liposomes were prepared by lipid film rehydration, functionalized with TAT (47-57) using wet chemistry, and characterized using transmission electron microscopy to both determine spherical diameter (~100 nm) and verify successful liposome formation. Liposomes were loaded with one of three commonly used antibiotics to treat these bacterial infections, including vancomycin, methicillin, and ampicillin. Bacterial growth curve studies were performed to evaluate the antibacterial effect of the liposomes. Also, the cytotoxicity of the liposomes was tested with astrocytes and endothelial cells. In addition to these studies, we also performed BBB studies using an *in vitro* model where astrocytes were seeded at the bottom of a 24 well plate, then inserts with endothelial cells were added to each well. After five days, the endothelial cell inserts were placed inside a separate 24 well plate containing bacteria. TAT-functionalized liposomes were added to see if they were able to cross the BBB and fight the bacteria located at the bottom of the plate.

Initial experiments were focused on the antibacterial properties of the TAT-functionalized liposomes. Their inhibitory effect on bacterial growth was tested with bacteria commonly associated with bacterial meningitis, including *Streptococcus pneumoniae*, methicillin-resistant *Staphylococcus aureus* (MRSA), and *Escherichia coli*. The results from these antibacterial studies demonstrated excellent growth inhibitory effects of TAT-functionalized liposomes loaded with methicillin for treating MRSA.

Cytotoxicity assays with endothelial cells and astrocytes demonstrated a notable increase in percent cell viability using functionalized and non-functionalized liposomes in comparison with free antibiotics.

Initial BBB *in vitro* studies showed that the readings of the transendothelial electrical resistance (TEER) decreased when the endothelial cells inserts were added to the 24 well plate with bacteria (Figure 1), indicating that the endothelial cells were dying. However, this decrease in TEER was notably less with the inserts containing our TAT-functionalized liposomes, and after the first hour the TEER values increased. This seems to indicate that our particles were able to cross the BBB and kill the bacteria located on the other side.

Figure 1. Blood-brain barrier static assay. TEER values obtained with TAT-functionalized liposomes loaded with methicillin (TAT-LipoMet) to fight MRSA

All results obtained for TAT-functionalized liposomes, including their low minimum inhibitory concentration (MIC) with respect to the use of free antibiotics plus their remarkable high percentage of cell viability with astrocytes and endothelial cells, provides evidence that this liposomal system can be a safe, alternative means for treating bacterial meningitis. Next steps in this research include additional BBB studies, utilizing static and dynamic BBB models, and investigating the effect of the liposome treatment on the genetic properties of the bacterial cell membrane, to better understand their cellular uptake mechanism.

This work was supported by funds from Northeastern University.

1. Pignatello, R., et al., Science Against Microbial Pathogens, 1, 52-60, 2011

2. Martinez-Gonzalez, R., et al. International Journal of Molecular Science Int. 17(8), 1209, 2016

3. Stalmans, S., et al., PLoS ONE, 10, pp. 1-22, 2015

538342 Uncorking and Oxidative Decomposition Dynamics of Gold Nanoparticle Corked Carbon Nanotube Cups for Drug Delivery Studied *Via in Situ* Transmission Electron Microscopy

Abstract:

Nitrogen-doped carbon nanotube cups (NCNCs) are an intriguing material for drug-delivery applications due to their cup-shaped morphology and their propensity for chemical modification. NCNCs can readily uptake small molecules, which can be effectively sealed within the inner cavity through the formation of nanoparticles (NP) on the open end of the cup. The resulting nanocapsules can then undergo oxidative biodegradation through biologically available peroxidase enzymes, [1] effectively releasing their loaded cargo. This mechanism has been applied for Au NP-corked NCNCs loaded with chemotherapeutic molecules for immunotherapy in melanoma models.[2] However, a better understanding of the mechanisms of cargo release â€" through uncorking and/or nanotube degradation â€" is crucial for improving their effectiveness as drug delivery platforms. The environmental transmission electron microscope (ETEM) is a powerful tool to directly investigate these questions, combining sub-nanoscale spatial resolution with the ability to image Au NP-NCNCs under similar gas, thermal, and liquid environments as their intended applications.

In this work, we present our *in situ* ETEM study of the dynamic oxidation-dependent uncorking and degradation of Au NP-NCNCs. NCNCs were synthesized using a liquid (a mixture of xylenes, acetonitrile, and ferrocene) injection chemical vapor deposition method. The resulting stacked NCNCs were separated into short stacked segments [3,4] and corked with Au NPs through a sodium citrate reduction of hydrogen tetrachloroauric acid. TEM specimens were prepared from aqueous solutions of the NCNCs. We used a Hitachi H9500 ETEM equipped a homebuilt multi-species gas injection system and specialized sample heating holders to study the structure of the Au NP-capped NCNCs and the dynamics of their uncorking and subsequent degradation *in situ*.

Structural examination of the NP-capped NCNCs revealed a multilayered tube structure, typically with multiple internal compartments. The Au NP morphology consisted of either a cap (interacting only with the uppermost lip of the NCNC) or plug (extending deeper into the tube). Under vacuum, the corked NCNCs exhibited remarkable tolerance to temperatures of up to 800 °C, though the internal structures of the carbon tubes exhibited restructuring at around 500 °C. The segmented cavities transformed from truncated cones in shape to spherical cavities, decreasing in apparent volume by 50-60%. This may be a result of increasing pressure on the walls from the encapsulated cargo as it is heated or due to thermally induced stresses. The Au NP caps were slowly "pulled" into the tubes at high temperatures (700-750 °C) while the Au plugs showed no meaningful change.

Upon exposure to ~10^{-2} Pa of O_2, however, the ingrown Au plugs were ejected from the tubes at a temperature-dependent rate that increased 2-3 orders of magnitude from 400 °C to 800 °C. This uncorking was of a stepwise, punctuated nature, rather than continuous. Importantly, the onset of uncorking occurred prior to any observed oxidative degradation of the NCNCs, significant decomposition of which required temperatures of ~500 °C. The exterior walls and any interior walls of cavities that were exposed (*e.g.*, from a breach) exhibited the same manner of attack. The oxidation primarily initiated at the rims (exterior wall) and cup bottoms (interior wall) of each of the NCNCs, progressing along the cup walls. Oxidation of the exterior walls typically proceeded at a slower rate and after a period of delay, attributed to the presence of an amorphous carbon layer on the outside of the tube, which could act as a sacrificial barrier hindering oxidation. In all cases, once oxidation began, it occurred along the entire length of the tube, with no preference for the ends. In contrast, interior cavities without an obvious breach exhibited an isotropic thinning of their walls, indicating a different mechanism was involved. This behavior is currently under investigation. It is important to note that the oxidative degradation of the NCNCs in aqueous solution may not necessarily proceed by the same mechanisms as these solid-gas reactions. For this reason, ETEM experiments are underway using an *in situ* liquid cell holder to study the uncorking and degradation reactions in the enzymatic solution environments. Surface chemistry changes are being explored by *in situ* X-ray photoelectron spectroscopy experiments to correlate with the gas-based ETEM structural changes. The knowledge gained from these studies will enhance the tailoring of the Au NP-corked NCNCs properties for a more effective therapeutic delivery.

[1] I Vlasova, *et al.*, Toxicology Applied Pharmacology **299** (2016), p. 58.

[2] Y Zhao, *et al.*, Journal of the American Chemical Society **137** (2015), p. 675.

[3] Y Tang, *et al.*, Journal of Physical Chemistry C **117** (2013), p. 25213.

[4] Y Zhao, *et al.*, ACS Nano **6** (2012), p. 6912.

534121 Length-Dependent Uptake, Inflammation, and Intracellular Processing of Single-Walled Carbon Nanotubes in Macrophages

Abstract:

Single-walled carbon nanotubes (SWCNTs) are increasingly being investigated for biomedical applications due to their chemical inertness, high surface area-to-volume ratio, and optical properties. Understanding cellular uptake and intracellular processing of SWCNTs while preserving their inherent properties is important for realizing SWCNT-based drug delivery and imaging systems. Especially, understanding immune-cell specific processing of SWCNTs is of great interest since immune cells involve in many physiological conditions including wound healing, inflammation, and cancer. Macrophages, a type of immune cell, have shown to take up larger number of SWCNTs than other cell types. They are professional phagocytes capable of two distinct size-dependent cellular entry mechanisms; endocytosis and phagocytosis. To determine how SWCNT lengths affect uptake and intracellular processing in macrophages, we non-covalently dispersed SWCNTs with average lengths of 50 nm (ultra-short, US) and 150 nm (short) with bovine serum albumin, which maintains SWCNT optical properties while promoting cellular uptake without affecting cell viability. Using confocal Raman imaging and spectroscopy, we quantify uptake, intracellular dispersion state, and recovery as a function of SWCNT lengths in macrophages. As expected from previous studies, macrophages uptake SWCNTs in dose-dependent manner for both short- and US- SWCNTs. However, uptake amount was significantly larger in a case for US-SWCNTs compared to short-SWCNTs. Interestingly, short-SWCNTs become highly bundled in phase dense regions of macrophages uptake and mostly retained for at least 24 h, whereas US-SWCNTs remain comparatively individualized and exocytosed over time. Phase contrast imaging of macrophages engulfed of SWCNTs revealed distinct phase dense regions, which co-registered with high SWCNT Raman signal. Further analysis of these regions revealed that they were mostly consisting of SWCNT bundles. In addition, we observed more numerous vacuoles and larger cell area for US-SWCNTs laden macrophages. This lead us to hypothesized that macrophages may have length-dependent inflammation upon exposure to SWCNTs, which may affect their uptake and intracellular trafficking in macrophages. The differential length-dependent inflammation and intracellular processing in macrophages may contribute to tailoring SWCNT properties for their usage in therapy and imaging. While high loading and subsequent release of individualized US-SWCNTs and triggering pro-inflammatory behavior in macrophages are more preferential for cancer therapy and drug delivery applications, high retention of short-SWCNTs in macrophages may be desirable for macrophage tracking *in vivo* and biomedical imaging applications.

524235 Nanoparticle Optimization for Improved Vaginal Drug Delivery during Pregnancy

Abstract:

Introduction: Vaginal drug delivery is the preferred route for targeting the female reproductive tract. By avoiding the harsh gastrointestinal environment and hepatic first-pass effect, local tissue concentration can be increased [1]. Additionally, the uterine-first-pass effect leads to accumulation of vaginally administered drugs in the uterus [2]. Vaginal formulations have been demonstrated to have some efficacy in prevention of preterm birth (PTB), though the optimal formulation for vaginal drug delivery has yet to be explored. Key barriers to effective vaginal drug delivery, including the protective mucus that acts as a barrier and facilitates clearance, must be considered [3]. By identifying physicochemical properties for optimal nanoparticle penetration of CVM, improved vaginal drug absorption can be achieved. Here, we characterized the structural and barrier properties of cervicovaginal mucus (CVM) during pregnancy using fluorescent probe nanoparticles of various sizes and surface coatings. We further utilize these design criteria to formulate drug nanoparticles that can penetrate the vaginal mucus barrier for more effective prevention of PTB.

Methods: 250 CVM samples (spaced monthly) were self-collected by 70 pregnant women. CVM microstructure was characterized by a quantitative multiple particle tracking method. Fluorescently-labeled nanoparticles with and without mucoinert coatings were added to CVM. Particle motions due to thermal energy were recorded and quantified to estimate the mesh spacing of mucin proteins comprising CVM. Trichostatin A (TSA), a histone deacetylase inhibitor that has been shown to delay parturition at term [4], and progesterone (P4), the â€œpregnancy hormoneâ€, were formulated as nanosuspensions (NS) via wet milling. To investigate the efficacy of these particles, pregnant mice were challenged with an intra-uterine (IU) lipopolysaccharide (LPS) injection on embryonic day 15 (E15, out of 19 day gestation). Sham animals received IU saline and no treatment. LPS animals received no treatment, TSA, P4, or TSA+P4. Treatment was dosed vaginally daily from E15-18. Mice were checked every 12 h for signs of PTB.

Results: Similar to what was observed previously in CVM from non-pregnant women, nanoparticles with mucoinert coatings have increased mobility in CVM from pregnant women. However, CVM from pregnant women causes a significant reduction in mobility of nanoparticles, particularly for nanoparticles 500 nm in size or greater, as pregnancy progresses. Mucoinert nanoparticles ~250 nm in size were slowed, but a significant fraction were still highly mobile in CVM. This inferred reduction in pore size in CVM is likely due to increasing progesterone levels throughout pregnancy. The TSA NS had average size of 207 nm with an average PDI of 0.25, and an average zeta potential of -0.38 mV. The P4 NS had average size of 301 nm with a PDI of 0.22, and an average zeta potential of -2.01 mV. These nanoparticles were shown to bypass the vaginal mucus barrier in mice, and have mobility in human CVM. They further showed efficacy in the model of inflammation induced PTB. The percentage of pregnant animals that went to full term were as follows: sham animals (100%), LPS only (10%), LPS + P4 NS (15%), LPS + TSA NS (30%), LPS + TSA/P4 NS (40%) ($p=0.003$ compared to LPS only). There was no significant loss of pups in litters that went to term. Analysis of cervical and myometrial tissues revealed anti-inflammatory effects from the combination dosing.

Conclusion: Here, we gained valuable insight for engineering targeted vaginal drug treatments for pregnant women. Further studies will use human myometrial and cervical biopsies to confirm the mechanism of action, and expand the delivery platform to other potential drugs for the prevention of PTB.

1. Ensign, R. Cone, J. Hanes, JCR 2014, 0:500-5:14
2. De Ziegler, C. Bulletti, B. De Monstier, A.S. Jääskeläinen, Ann N Y Acad Sci, 1997; 828:291-9.
3. Yu, J. Chisholm, W.J. Choi, A. Anonuevo, S. Pulicare, W. Zhong, M. Chen, C. Fridley, S.K. Lai, L.M. Ensign, J.S. Suk, J. Hanes, Adv Healthc Mater, 5 (2016) 2745-2750.
4. C. Condon, P. Jeyasuria, J.M. Faust, J.W. Wilson, C.R. Mendelson, PNAS 2003, 100.16, 9518-9523.,

532320 Responsive Foams for Nanoparticle Delivery

Abstract:

Responsive foams for nanoparticle delivery

Chang Tian[1], Christina Tang[2], Antoinette Nelson[3], Jennifer Holloway[3], Patrick J. Sinko[3], Robert K. Prudâ€™homme[1]

[1] Princeton University, Department of Chemical and Biological Engineering, Princeton, NJ, 08540

[2] Virginia Commonwealth University, Department of Chemical and Life Science Engineering,

Richmond, VA 23284

[3] Rutgers, The State University of New Jersey, School of Pharmacy, Piscataway Township, NJ 08854

Foams are attractive vehicles to deliver nanoparticle therapeutics since they can distribute nanoparticles over a large surface area with minimal liquid residue. We have developed two responsive foams to achieve rectal delivery of nanoparticles. Those foams are easy to apply, stable and room temperature and can be engineered to break in response to temperature or moisture. The temperature-responsive foams form and break based on the Pickering emulsion and phase transition of long chain alcohols. The moisture-responsive foam expands due to a reaction between sodium bicarbonate and citric acid, and breaks based on a surfactant dilution and destabilization of foam interfaces. Various *in vivo* experiments were performed to examine rectal delivery of nanoparticles by those foams. We encapsulated near-IR dyes inside the nanoparticles through Flash NanoPrecipitation (FNP) to record the coverage of nanoparticles inside mouse colon, using FX PRO molecular imaging system. In addition, introducing lanthanides inside nanoparticles provides a way to monitor nanoparticle retention in the GI tract by ICP-MS.

527474 Effect of Hydrocarbon Contamination on the Electrochemical Activity and Double Layer Capacitance of Graphitic Carbons

Abstract:

For nearly eighty years, graphitic carbons have been considered as hydrophobic. However, recent works in our labs showed that graphite and graphene are much more hydrophilic than previously thought. Since the water-graphitic interface is critical to many important applications including electrodes, adsorbents, catalyst support, and solid lubricants, this unexpected finding could completely change the way graphitic materials are made, modeled and modified. In this presentation, we will discuss on the intrinsic water wettability of graphitic carbons and its implications. First, our recent experimental results, e.g., contact angle, XPS, ATR-FTIR and ellipsometry, will be presented showing that pristine graphene and graphite are mildly hydrophilic and airborne hydrocarbons make the surfaces appear to be hydrophobic. Second, these results are further rationalized by analyzing the change in surface energy of the graphitic surfaces before and after hydrocarbon contamination. Third, implications of contamination will be illustrated through comparing the electrochemical activity and double layer capacitance of pristine and contaminated graphite. Lastly, future research directions to address a few important unanswered questions will be presented.

528010 Colloidal Electronic Cells Based on 2D Materials

Abstract:

Graphene and other 2D materials possess desirable mechanical and functional properties for incorporation into or onto novel colloidal particles, potentially granting them unique electronic and optical functions. However, this application has not yet been realized because conventional top-down lithography scales poorly for the production of colloidal solutions. Herein, we develop an â€œ*autoperforation*â€ technology providing a means of spontaneous assembly for colloidal microparticles comprised of 2D molecular surfaces at scale. Such particles demonstrate remarkable chemical, mechanical and thermal stability. They can function as aerosolizable and dispersible electronic particles capable of storing and transferring digital information, as well as recoverable microprobes for sensing and recording chemical information in water and soil.

523922 One-Step Non-Distructive Decoration of Transition Metal Oxide Nanoparticles on Large Scale Graphene for Electronic and Sensing Applications

Abstract:

Binding graphene with transition metal oxide nanoparticles(TMON) is widely used in the fields of gas sensing, electrochemical catalysis and supercapacitors. However, it is still challenging to incorporate nanoparticles on pristine graphene, since graphene is chemically inert and has no nucleation/attachment sites. Here, we show a novel and facile approach to deposit TMON (Cr_2O_3/CrO_3, MoO_3 and WO_3) on large-scale graphene by one-step photochemical reaction. The X-ray photoelectron spectroscopy shows that C≡O group is introduced after the attachment of TMON on graphene, suggesting that graphene is first functionalized based on η^6 chemistry. Further, TMON formed in solution are attached on η^6-functionalized graphene. Based on the Raman spectrum, the decoration of TMON with a size ranging from 20 to 50 nm leads to a blue shift in both G and 2D band, suggesting p-doping of graphene. Also, the electronic transport measurements are performed on back-gated field effect transistor. It shows that the Fermi level is moved downward by 250 mV, with Dirac point shifting to 50 V after one-step reaction. TMON@graphene on n-silicon is applied as Schottky junction solar cells, which has an enhanced power conversion efficiency due to its high erelectrical conductivity and larger work function.

537715 Large-Scale Chemical Synthesis of Graphene for Energy Storage and Biological Applications

Abstract:

Interest for two-dimensional (2D) nanomaterials such as graphene primarily stems from their unique intrinsic properties, combined with multifunctionality. Graphene exhibits fascinating physical and chemical properties that can be controlled through precise engineering processes. As a result, graphene has found numerous applications in areas such as polymer composites, energy storage, or bioengineering. In the first part of this talk, I will describe the large-scale synthesis of graphene obtained through the oxidation of graphite, followed with a description of the fundamental physical and chemical characteristics of graphene. In the second part, I will explore the promising applications of graphene in the area of energy storage and biology. In particular, I will first analyze and discuss the role of graphene as a component of Li-ion batteries to enhance their performance. Then, I will subsequently cover another example of effective contribution of graphene to the area of biological materials.

538466 Achieving High Open-Circuit Voltage in Graphene/Silicon Photovoltaic Cells with h-BN Tunneling Layer

Abstract:

Graphene-on-Silicon (G/Si) Solar Cell has attracted numerous interest due to its potential to be an alternative to the Si wafer-based photovoltaic technologies. However, such heterojunction photovoltaic cells suffer from long-term instability in power conversion efficiency. This is mainly due to the interfacial charge carrier recombination owing to the low G/Si Schottky barrier height. Therefore, it is critical to overcome these challenges *via* interfacial engineering. Here we have introduced two-dimensional (2D) hexagonal boron nitride (h-BN) crystal as a tunneling inter-layer in graphene/Si heterojunctions. Thin films of h-BN is directly synthesized on lightly-doped Si surface *via* bottom-up chemical-surface-adsorption strategy followed by the transfer of chemical vapor deposited graphene. The 2D/2D/3D architecture of Graphene/h-BN/Silicon forms a metal-insulator-semiconductor (MIS)-type junction, where h-BN acts as an electron-blocking or hole-transporting medium and it avoids interfacial charge carrier recombination. An increased open-circuit voltage (V_{OC}) is found and this may be due to the decrease in interfacial charge carrier recombination and increase in light generated current.

538588 Multicolumn Gel Chromatography for Scalable Separation of Nanosheets Based on Size and Surface Chemistry

Abstract:

Liquid-phase exfoliation of two-dimensional nanomaterials yields nanosheet dispersions with broad size distribution and varying surface chemistry. This polydispersity in size and surface chemistry of nanosheets becomes less desirable when their properties are to be precisely controlled for different applications. Hence, scalable separation methods are highly demanded to make this class of materials industrially relevant. Here, we use multicolumn gel chromatography to separate nanosheets of graphene family based on their size (later size and thickness) and surface chemistry (density of functional groups or defects). We demonstrate that various parameters such as stabilizer type, pH, and ionic strength of the solvent alter the binding affinity of nanosheets on the gel and tune the elution time of nanosheets of different size and surface chemistry. Single particle tracking was used as a powerful tool for quick evaluation of nanosheets size distribution in the bulk. Other spectroscopic techniques such as fluorescence and Raman spectroscopy were applied to investigate the surface of the separated nanosheets.

538620 Graphene Interfaced *Geobacter* for Improved Electron-Transport Channels in Microbial Fuel Cell: A Single Cell Investigation

Abstract:

With the increased interest in developing cost-effective and readily available energy sources, the research in the microbial fuel cell (MFC) field has gained an increased attention. Here, we report the use of graphene's high electron mobility coupled with *Geobacter sulfurreducens*' biocatalyst nature to increase the current response of a MFC. It is well-accepted that one of the limiting factors of current generation of a *Geobacter* based MFC is the electron transport process in the anodic chamber. We addressed this limitation by utilizing the enhanced electrical conductivity of graphene. In our preliminary tests, the current density of the MFC increased by 9 to 13 folds by interfacing *Geobacter* cells with graphene. To study the interface between graphene and *Geobacter*, Raman spectroscopy and scanning electron microscopy (SEM) were employed. The Raman characterization studies show that a single geobacter interfaced with graphene n-dopes it by about $\sim 3 \times 10^{12}$ cm^{-2}. These results demonstrate that the graphene-*Geobacter* interface provides an approach to overcoming the electron-transfer related challenges of conventional MFCs.

524128 Synthesis of Graphene from Biochar

Abstract:

Graphene, "the wonder material" was discovered in 2004. It consists of one atom thick carbon atoms which is 100 times stronger than steel and conducts electricity. Potential applications of graphene include membranes for water purification, composites and coatings, in batteries for energy storage, microprocessors, sensors etc. Even though graphene has a huge potential, production costs limits its use. In this research, we use biochar, a solid carbon-rich material produced from biomass model compounds (cellulose, hemicellulose and lignin) produced by hydrothermal carbonization (300 ºC and 9.0 MPa), as the raw material. The biochar was further treated with 0.1 N HNO_3 to remove any soluble alkali and alkaline earth metal ions and bio-crude residues. The biochar was grinded (<0.17 mm) and uniformly distributed over copper and iron foils and heated to a temperature of 1000 ºC for 2 hours in a quartz tube with constant N_2 flow. The uniqueness of this approach is that it doesn't require vacuum as in chemical vapor deposition to deposit graphene on metal foils. Our reactor is a simple quartz tube heated in a furnace. The role of H_2 was also studied by mixing it with the N_2 gas (1:3 vol. basis). The metal foils were analyzed by Scanning Electron Microscope (SEM), X-Ray Photoelectron spectroscopy (XPS), and Raman spectroscopy for characterizing the graphene. Graphene was observed on iron foils while no graphene growth was observed on the copper foils.

538737 Stabilizing Phosphorene *Via* Hexagonal Boron Nitride Passivation

Abstract:

Layered crystals have revolutionized the field of nanoelectronics and optoelectronics due to their potential to be exfoliated (mechanically or chemically) into atomically-thin two-dimensional (2D) surfaces (for example graphene from graphite). Several 2D crystals and their complex van der Waals heterostructures are recently realized *via* micromechanical cleavage (peeling) technique. Phosphorene, a two-atom thick 2D material has found significant attention because of its unique properties including: sizable band gap (0.3-1.2 eV), in-plane anisotropy, and high charge carrier mobility. Further the 2D phosphorene layers can be easily exfoliated from the bulk 3D black phosphorous crystals and transferred onto any arbitrary substrates. However, the major problem associated with the technological applications of phosphorene is that these 2D material crystal structures can deteriorate in the ambient condition (in water or oxygen). Here, we stabilize the crystals of phosphorene by passivating with a 2D layer of hexagonal boron nitride (h-BN). Heterostructures of phosphorene and h-BN (h-BN/phosphorene/h-BN) were fabricated to remove lattice degradation of phosphorene from both the top and bottom surfaces. Confocal Raman vibrational spectroscopic results indicate that the h-BN/phosphorene/h-BN architecture (capsulated phosphorene) is more stable in contrast to the un-capsulated phosphorene (phosphorene/h-BN or phosphorene/SiO2). Interestingly, we observe that the structural degradation nucleates from the basal surfaces rather than edges or steps.

533546 Conformations of Weak Polyelectrolytes in Confined Geometries

Abstract:

Recently designed charge-rejection and mosaic membranes hold exceptional promise for low-power water purification and material reclamation. The use of polyelectrolyte brushes and oligomeric grafts is key to controlling the behavior of these systems. Importantly, the surface charge on the interior of these membranes is controlled by the interplay between adhesive molecular interactions, pH, salt content, and polymer density. I will discuss the use of coarse-grained constant-pH molecular dynamics simulations for simulating the charging equilibrium and conformational changes in weak polyelectrolytes, with an explicit focus on confined polyelectrolyte chains and brushes. We demonstrate that the increased monomer densities and reduced chain entropy present in confinement can have significant effects on the apparent dissociation constant of a weak acid groups in a polymer chain.

528016 Quantifying Structure-Function Relationships of Protein-Selective Networks at the Micro- and Macro-Scale

Abstract:

Polymeric formulations and structures that act as high-affinity protein sorbents have a wide range of industrial and pharmaceutic applications. For example, high-affinity sorbents will aid in the separation of active therapeutic proteins from bioprocess byproducts via chromatography[1], but can also be applied in capsules and carriers to enhance the therapeutic's loading capacity[2]. Herein, we developed libraries of polymeric networks for high-affinity binding and maximal adsorption capacity of model cationic proteins. Simultaneous pursuit of affinity and loading is particularly interesting because the polymer properties that enhance each are paradoxical. By example, while hydrophobic moieties generally enhance protein-material affinity by enabling the formation of entropically favorable hydrophobic interactions, hydrophobicity decreases the polymer-solvent interaction parameter governing swelling (and loading capacity by association) in aqueous media. In this study, we present the synthesis and assessment of networks that possess a variety of composition and macromolecular structure, to understand this complex interplay and engineer useful protein sorbents.

The base network structure for all studies was a copolymer of methacrylic acid (30 mol%) methylenebisacrylamide (5 mol%) and acrylamide (remaining mol fractions). These networks were copolymerized with up to 40 mol% of either methyl methacrylate (MMA), tert-butyl methacrylate (tBMA), or benzyl methacrylate (BzMA) in substitution for some, but not all, of the acrylamide fraction. Microparticles of the above formulations were synthesized as films by UV-initiated free radical polymerization using Irgacure 184 as a photoinitiator, and manually crushed/sieved to yield a fine powder (<45 μm dry diameter). Nanoparticles of poly(acrylamide-co-methacrylic acid), which were necessary for many analyses, were synthesized by inverse emulsion polymerization as previously described[3]. For peptide-modification studies, pendant methacrylic acid moieties within nanoparticles were modified post-synthesis with a bifunctional amine-maleimide linker and a cysteine-containing peptide via sequential carbodiimide-coupling and thiol-maleimide Michael addition reactions. The composition of all copolymers were verified by FTIR, potentiometric titration (for methacrylic acid content) and a MicroBCA colorimetric assay (as appropriate, for peptide content). The morphology of microparticles was determined by electron microscopy, and the diameter of nanomaterials was determined by DLS.

Microparticles with a range of hydrophobic content were incubated individually with model proteins (i.e. lysozyme, cytochrome c, trypsin, bovine serum albumin, bovine hemoglobin, ovomuccoid) that possessed a range of molecular weight and isoelectric point. Adsorption of each species correlated strongly with the formation of ionic polymer-protein interactions, as evidenced by the substantial adsorption competition in high ionic strength buffers. Hydrophobic interactions, within certain constraints (up to 30% MMA, up to 20% tBMA, up to 10% BzMA) enhanced the adsorption of all high isoelectric point proteins to the microparticles, corresponding with the point of significant reduction in water content (equilibrium swelling studies). Hydrophobicity played a particularly major role in lysozyme adsorption, with optimal formulations loading 6.87-fold (30% MMA), 7.50-fold (10% tBMA), and 7.18-fold (5% BzMA) more lysozyme at saturation than formulations without hydrophobic monomer. Analysis of the solvent-accessible surface area (SASA) amino acid composition of each model protein lead to the conclusion that arginine and tryptophan residues (prevalent in lysozyme but not trypsin or cytochrome c) may be primarily responsible for lysozyme's sensitivity to hydrophobic interactions, even in the presence of physiologic salt.

Protein-microparticle interaction studies, therefore, pointed to that small quantities (e.g. just 5 mol% of benzyl methacrylate) of hydrophobic moieties can optimize the affinity-permeability tradeoff and maximize protein partitioning while minimizing macroscopic effects on the network (i.e. water content, elastic modulus). To the aim of capitalizing on this optimum, while simultaneously enhancing single-protein network specificity, we conjugated hexamer peptides (identified previously[4,5]) to poly(methacrylic acid-co-acrylamide) networks. These hexamers were optimized for trypsin adsorption, and led to enhancement of trypsin affinity, relative to unmodified controls as evidenced by equilibrium protein adsorption in quartz crystal microbalance with dissipation (QCM-D) studies.

In summary, an affinity-permeability tradeoff exists for hydrogels that contain hydrophobic moieties, as the formation of hydrophobic protein-polymer interactions and interaction of moiety-containing chains with solvent (water) must balance. Optimal levels of each hydrophobic monomer tested (MMA, tBMA, BzMA) were determined that did not deter hydrogel swelling in 1x phosphate buffered saline and enabled up to 7.5-fold enhancement in lysozyme adsorption. Further protein-specific adsorption enhancement was also explored for trypsin-peptide nanogel conjugates, following the affinity-permeability tradeoff guidelines established in microparticle studies. This study not only developed promising sorbents for lysozyme and trypsin, but through identifying a relevant thermodynamic tradeoff in gels will also inform the fabrication of high-affinity sorbents for alternate therapeutic proteins.

References:

[1] Xiu, L., Valeja, S. G., Alpert, A. J., Jin, S., & Ge, Y. (2014). Effective protein separation by coupling hydrophobic interaction and reverse phase chromatography for top-down proteomics. ***Analytical chemistry***, ***86***(15), 7899-7906.

[2] Culver, H. R., Clegg, J. R., & Peppas, N. A. (2017). Analyte-responsive hydrogels: intelligent materials for biosensing and drug delivery. ***Accounts of chemical research***, ***50***(2), 170-178.

[3] Zhong, J. X., Clegg, J. R., Ander, E. W., & Peppas, N. A. (2018). Tunable Poly ((methacrylic acid) co acrylamide) Nanoparticles through Inverse Emulsion Polymerization. ***Journal of Biomedical Materials Research Part A***.

[4] Clegg JR, Gu J, Harger, M. and Peppas NA. Molecularly Imprinted Polymer-Peptide Hybrid Materials for the Recognition and Sequestration of Proteins. Biomedical Engineering Society Annual Meeting, Phoenix AZ. 10/14/17

[5] Clegg JR, Zhong JX, Ander EW, Sun JA, and Peppas NA. Versatile Anionic Nanogels for the Targeted Delivery of Therapeutic Small Molecules via Active and Passive Mechanisms. US-Japan Drug Delivery Symposium, Maui, HI. 12/16/17.

538875 Block Copolymer Directed Self-Assembly Using Chemoepitaxial Guiding Underlayers with Topography

Abstract:

Simulations have been used extensively to analyze the behavior of block copolymers (BCPs) in directed self assembly (DSA) to produce nanoscale features in polymer photoresists. Typically these simulations have targeted patterns of specific chemical makeup (chemoepitaxy), or three dimensional lines (graphoepitaxy). However many applications of these patterns contain both chemoepitaxial and graphoepitaxial features. Recently, hybrid processes for BCP DSA employing both chemoepitaxy and graphoepitaxy have been developed that use underlayer patterns to induce DSA of BCPs. However, little is understood about the correlated effects of both chemoepitaxy and graphoepitaxy in DSA We have carried out coarse-grained molecular dynamics simulations of BCPs using a variety of hybrid underlayer patterns to characterize DSA as a function of these geometries. Simulations of various hybrid DSA approaches currently in use were simulated. These included the Liu-Nealy (LiNE), Surface Modification for Advanced Resolution Technology (SMART), and coordinated line epitaxy (COOL) process flows currently being used in BCP DSA.

It was found that underlayers with vertical sidewalls behave in manners similar to purely graphoepitaxial guiding underlayers, while underlayers with sloped sidewalls behave in a manner similar to chemoepitaxial guiding underlayers. With vertical sidewalls, it is found that larger topographic step heights decrease pattern defectivity, though with diminishing returns. The width of the trench should be an integer multiple of the natural repeat distance of the block copolymer, though the width of the trench can have approximately 10% variation before significantly affecting the defectivity. These results suggest that the width of the guiding regions should be spread out in order to reduce defectivity. It was found that the chemical preference of both the top of the mesa and bottom of the trench have little effect on the defectivity so long as the preference is sufficienty neutral to prevent the formation of horizontal lamellae. A variety of sloped sidewall geometies were explored. Sidewalls that undercut the mesa were found to increase defectivity. It was found that certain trapezoidal shaped topographies can reduce defectivity, especially when the top of the mesa, as well as the width of the sidewalls, were approximately the width of one lamellae. In these cases we observed an optimal height of the topography. The optimal height appears to be the height that makes the effective width of the sidewall the same width as a lamellae. Additionally, it was found that a slightly preferential bottom of the trench and a very preferential top of the mesa helped further lower defectivity. An additional sloped sidewall case, where the topographic feature was triangular in shape was explored. It was found that this triangular topographic feature yields far lower defectivity than a similarly sized rectangular topographic feature.

529675 A New Class of "Gecko Leg" Dendrimeric Polymeric Particles By Interfacial Templating of Multiphasic Liquids

Abstract:

Making mechanically stable coatings and bulk materials from polymer microparticles typically requires the use of a binding agent. An alternative strategy for glue-free adhesion and cohesion could be “contact splitting” which maximizes the contact area of an adherent to a substrate, as exemplified in gecko lizards' nanofibrous setae. We will present a new class of nanofibrillated dendrimeric polymer particles (DPPs), which exhibit contact splitting effects on the microscale and are highly adhesive and cohesive due to their morphology. The DPPs are fabricated by an efficient and scalable process for liquid-based synthesis of nanomaterials by antisolvent precipitation in turbulently sheared medium. The process allows a variety of polymers to be readily shaped into nanofibrillated particles. The DPPs are hierarchically structured, with a big branched corona of nanofibers spreading out in all directions. The biomimetic similarity of their structure to the gecko lizards' setae endows DPPs with excellent adhesion and cohesion properties. Our results demonstrate that this strong adhesion and cohesion can be attributed to the contact splitting phenomena and van der Waals interactions of their nanofibrous structures. This novel class of polymeric particles offers means to make strong non-covalent binding coatings and new types of dry adhesives. In addition, the DPPs exhibit high excluded volume, stemming from their hierarchical structure. They can build strong three-dimensional networks in numerous liquids, leading to gel-like behavior at concentrations as low as 1-2 vol.%. These particles have a broad range of potential applications and represent a system of fascinating phase behavior and structure building ability.

536660 Effect of Asymmetric Homopolymer Addition on Structural Characteristic of Lamellae Forming Block Copolymers Aligned Via Directed Self-Assembly

Abstract:

Block copolymers (BCPs) offer an attractive pathway to the formation of nanoscale patterned features in semiconductor integrated circuit manufacturing and other applications due to their inherent propensity to microphase separate into morphologies (lamellae, cylinders, spheres, etc.) with feature pitches in the range of approximately 5 nm to 100 nm. While the native microstructures formed in such block copolymer microphase-separated assemblies are generally highly defective due to entropic effects, it has been shown that directed self-assembly (DSA) techniques (e.g. graphoepitaxy and chemoepitaxy) which rely on the formation of larger â€œguide patternsâ€ using often traditional lithographic techniques at larger length scales can provide an energetic landscape on surfaces in which the location and orientation of the microphase separated regions formed from the block copolymer can be controlled. These controlled, essentially templated, micro-phase separation processes can produce structures with extremely low numbers of defects in nearly perfectly ordered morphologies over large area. Furthermore, the sub-division of the larger guiding structure pitch into much smaller pitch structures through the block copolymer micro-phase separation and self-assembly process is the critical advantage of this process which results from combining a top-down lithographic patterning method with a self-assembly process. It is exactly this ability to form virtually perfect nanometer scale assemblies over large areas that presents the real opportunity for DSA of block copolymers to extend and enhance the current top-down lithographic capabilities engineered by the semiconductor community over the last six decades.

In engineering the materials used for such DSA patterning methods, the pitch of the features formed is controlled not by the lithographic techniques used to form the guide patterns, but instead are determined by the molecular weights of the blocks formed in the copolymers. This allows synthetic control over feature sizes which can be tightly determined using for example modern living polymerization methods, and this ultimately is the critical advantage that allows DSA methods to surpass the diffraction limits of state-of-the-art lithographic processes. However, this also means that any small desired variation in feature size using DSA of pure block copolymers must be accomplished by synthesizing new block copolymers targeted for that new feature size. It also means that exquisite molecular weight control is required to yield useful batches of copolymer targeted at specific pattern feature sizes as any variation in molecular weight of the block copolymer directly translates into feature size changes in the resulting assembled block copolymer film pattern. It is therefore of interest to explore methods whereby adjustments or variations in the pattern (i.e. morphology) pitch that can be achieved with a given block copolymer can be made. Two such methods that are straightforward to imagine are: (1) addition of homopolymer of one or more blocks in the copolymer and (2) blending of block copolymers of different molecular weight. The first methods offers the advantage that in principle it should be possible to vary somewhat independently the size of the different domains in the resulting morphology (e.g. the size of the two different lamellae domains forms in a lamellae forming diblock copolymer could be independently adjusted while the overall pitch of the pattern is also modified through addition of different amounts of homopolymer of the two blocks constituting the block copolymer). However, the effect of the addition of such homopolymer additives is not obvious given the different types of interactions and constraints in the material system for homopolymer as compared to the block copolymer. The goal of the work reported here is to elucidate the effect of such homopolymer addition on the lithographically relevant characteristics of block copolymer DSA processes.

Addition of homopolymer is already known to increase the achievable pitch of block copolymer materials, but the details of this process have not been deeply investigated. Furthermore, patterned features in integrated circuits and other possible applications are not all spaced by distances equal to the feature size of interest (i.e. not all patterns are so-called 1:1 patterns), and thus may exist at more complex geometries and pattern size ratios than allowed by standard di-block copolymer self-assembly may allow. The addition of homopolymer to the block copolymer not only allows for an increase in the overall pitch of the repeating morphology, but the addition of different amounts of the homopolymer constituents of the block copolymer can allow for variation in the domain sizes of a morphology (e.g. it is possible to achieve lamellae patterns where the lamellae widths are not equal even though a 50:50 volume fraction diblock copolymer is used for the assembly). This allows for formation of non-equal block copolymer pattern domain feature sizes in morphologies not achievable by the block copolymer alone (e.g. it may be possible to go to different domain size ratios in a lamellae forming diblock copolymer) that if attempted from a single diblock copolymer would result in phase transition to different morphologies such as transitioning from lamellae to cylinder phases when varying to far from 50:50 volume fraction blocks. In this work, coarse grained molecular dynamics simulations have been used to probe the effect that asymmetric homopolymer addition has important DSA patterning behaviors and pattern characteristics. For example, the effect of homopolymer addition on the line edge- and line width roughness (LER/LWR) of lamellae forming BCPs has been probed as a function of both homopolymer to block copolymer loading and as a function of varying the ratio of the two homopolymers that are constituents of the block copolymer. The limits of such processes, in the form of effective process windows, will be presented and the general trade-offs presented by such processes will be discussed.

524512 Theory and Simulation Studies of Structure and Thermodynamics in Polymer Nanocomposites Containing Grafted Nanoparticles

Abstract:

Polymer nanocomposites (PNCs) are a class of polymeric materials where nanoparticles are added to a polymer matrix to achieve desired macroscopic properties. To achieve the desired macroscopic properties, one needs to control the PNC structure/morphology, i.e. arrangement of nanoparticles within the matrix. It has been shown that by tailoring the polymers that are grafted on the nanoparticles one can tune the effective inter-particle interactions and as a result, the PNC morphology. In this talk, I will share our recent work focused on one specific set of polymer chemistries where hydrogen bonding interactions are present in the graft and matrix chains. Having acceptor-donor (A-D) sites on graft and matrix polymers offers a promising means to obtain a precisely tuned morphology depending on the valency and strength of directional interactions. I will first present a coarse-grained model that we developed to mimic such directional and specific interactions between donor and acceptor sites on monomers along the graft and matrix. Using this model and molecular dynamics (MD) simulations we contrast the effect of isotropic and directional interactions between the graft and matrix chains on the PNC morphology. We characterize the PNC morphology, in particular, the wetting of the grafted layer by the matrix chains, through graft and matrix chain conformations and monomer concentration profiles. In this talk I will present the highlights of our study conducted over a large design space including graft and matrix chain lengths and flexibility, strength of isotropic and directional interaction, grafting density, etc.

522348 Computational Characterization of Ultrathin Amorphous Polymer Films in Liquids

Abstract:

Polymer materials have a wide variety of important applications in chemical processes. The ability to accurately predict the fundamental properties of polymers is the key to their optimization and rational design of new polymers. In this study, the material properties of an amorphous polymer in liquids are computationally characterized from a molecular perspective. In order to achieve this objective, molecular simulations and defined characteristic lengths are integrated to analyse the dynamic and equilibrium properties of amorphous polymer in liquids. The polymer of intrinsic microporosity (PIM-1) is used as an illustrative polymer as it has shown great potential in various gas and liquid separations. Molecular simulations of PIM-1 swelling in liquids such as organic solvents and water are explored. Solvent-induced swelling of PIM-1 is observed in organic solvents while the polymer structure remains intact in water. In view of the glassy nature of PIM-1, the polymer swelling is attributed to both solvent occupation of free volume and dilation of polymer matrix. With defined characteristic lengths, swelling process has been validated against a predictive model. In addition, gravimetric swelling degree of PIM-1 in solvents and their correlation with solubility parameters are in good agreement with experimental findings. Detailed analyses on the equilibrium structures of swollen PIM-1 offer further molecular insights. Lastly, the performance of PIM-1 as membrane material in seawater is assessed through the same characterization procedure. This study illustrates the usefulness of the characterization procedure in estimating material properties of PIM-1 in liquids. This procedure can potentially be applied to other polymers in liquids to gain fundamental understanding of their performances as membranes in liquid separations such as reverse osmosis and organic solvent nanofiltration.

536389 Photocrosslinking to Obtain Graphitic Carbon-Based Nanowires from Ordered Polymer Networks

Abstract:

Polyethylene oxide (PEO) has been widely used for electrospun microfibers due to its viscoelastic properties. However, PEO is not a suitable candidate for the production of graphitic nanowires from pyrolysis. In contrast, aromatic epoxy oligomers yield carbon-based materials with a higher content of graphitic carbon after pyrolysis. Here, we analyze the Near Field Electrospinning (NFES) of a mixture of PEO and multi-functional epoxy crosslinkers as a candidate for the production of spatiotemporally fixed polymer fibers that can be converted into conductive carbon-based materials. The fabrication of photopolymerizable micro- and nanofibers from NFES has remained relatively unexplored. Here, we explore the influence of the crosslinking reaction on the properties of the fibers and of the final nanowires. We correlate fiber diameter before and after irradiation to the viscosity, conductivity, and surface tension of the solutions as a function of the PEO loading using dimensional analysis. The cationic photopolymerization kinetics were analyzed to determine curing times and the effect of post-curing treatment. An ideal crosslinker/PEO ratio is proposed, which allows electrospinning at low voltages while ensuring the smallest fiber diameter possible, sub-micrometer range. While the physicochemical properties of SU-8/PEO solutions vary only slightly, the fiber diameter is significantly affected by the SU-8/PEO ratio. The correlation of diameter to PEO-content is based on SEM analysis of the electrospun fibers before and after pyrolysis, and fits well with previous studies of electrospun PEO-based microfibers. We report the first study on the effect of crosslinker/PEO ratio on final diameter for photopolymerizable PEO-based microfibers at the low voltages and short distances used in Near Field Electrospinning (NFES) and Electromechanical Spinning (EMS). These fibers were pyrolyzed to produce graphitic carbon-based micro- and nanowires useful for biosensors.

524684 Multi-Scale Simulations of the Fabrication of Polymeric Nanoparticles through Rapid Solvent Exchange

Abstract:

Structured nanoparticles are important for various scientific and industrial applications. A fabrication process which is simple, scalable and allows for precise control over particle size and morphology is thus highly desirable. Here, we describe a multi-scale simulation approach performed to study the rapid mixing of polymers in solution with a miscible non-solvent, *i.e.*, Flash NanoPrecipitation (FNP).[1] Using a combination of molecular dynamics (MD) and kinetic Monte Carlo (KMC) simulations, we studied the process on both nanoscale and macroscopic scale. On the nanoscale, our MD simulations revealed how nanoparticles with a range of surface structures, including Janus, core-shell and patchy, *etc.*,[2-3] arise from different feed materials. The simulations also indicated that nanoparticle morphologies and other properties such as the composition and patchiness can be reliably tuned. On the macroscopic scale, we performed KMC simulations on macroscopic length- and time-scales and studied how nanoparticle size and size distribution can be controlled by processing parameters specific to FNP, such as the mixing rate and the feed concentration. Our model produced results in agreement with the experiments and also allowed us to gain understanding on particle aggregation mechanism through the solvent displacement process. Our multi-scale simulation approach allows us to make predictions for the FNP process from a given set of feed polymers and processing parameters. Our results demonstrate that the process is highly promising for the production of structured nanoparticles with various surface features in a scalable and controlled way.

References:

Zhang, C, Pansare, V. J., Prudâ€™homme R. K. and Priestley R. D. **Flash Nanoprecipitation of Polystyrene Nanoparticles.** ***Soft Matter***, 2012, **8**, 86-93

Li, N., Panagiotopoulos, A. Z. and Nikoubashman, A. Structured Nanoparticles from the Self-Assembly of Polymer Blends through Rapid Solvent Exchange. *Langmuir,* 2017, 33, 24, 6021-6028

Li, N., Nikoubashman, A. and Panagiotopoulos, A. Z., Controlled production of patchy particles from the combined effects of nanoprecipitation and vitrification. ***Soft Matter***, 2017, **13**, 8433-8441

530956 Invited Speaker: Scalable Manufacturing Methods for Polymeric Nanoparticle Drug Delivery Systems

Abstract:

Although manufacturing technologies for hydrophilic drug carriers, such as solid polymer nanoparticles and liposomes have been well-established, there are fewer approaches for the delivery of hydrophobic drugs. Hydrophobic drug delivery represents a significant bottleneck in therapeutic development, especially for central nervous system therapeutics, which must also cross the blood-brain-barrier (BBB). It is estimated that 98% of small molecule drugs cannot cross BBB, resulting in the failure of 90% of molecules in development by the pharmaceutical industry (1). Thus, additional carriers for hydrophobic drug delivery are critically needed.

Micelle drug delivery carriers offer an attractive option for the solubilization of hydrophobic drugs. However, to be clinically effective manufacturing technologies for micelles should: produce product at high throughput, with high purity and minimal residual solvent, with high encapsulation efficiency, and efficient biodistribution and release at the target site. We are investigating scalable nanomanufacturing approaches for polymeric micelle synthesis, including flash nanoprecipitation (FNP) (2) and electrospray assisted-micelle formation (Aero-IS) (3) that scale-up batch water addition and interfacial instability approaches.

Here, we will discuss nanoparticle carrier design to optimize encapsulation of drugs with differing chemical structure (e.g., dexamethasone steroids, SAHA/vorinostat anti-cancer compounds, and lutein nutraceuticals). Chemical composition strongly influences encapsulation efficiency, but also plays a critical role in the surface properties of the resultant micelle. We will also evaluate competing scalable technologies (i.e., FNP vs. Aero-IS) discussing challenges in solvent removal, drug encapsulation efficiency, and final micelle structure that result from thermodynamic and kinetic limitations of these processes. Finally, we will discuss our efforts to design a drug delivery system for a model therapeutic, SAHA, for the treatment of glioma brain cancers, necessitating delivery across BBB.

References

1. Vlieghe et al., Medicinal Research Reviews, 33(3): 457, 2013.
2. Johnson et al., Physical Review Letters, 91(11): 118302, 2003.
3. Duong et al., Langmuir, 30(14): 3939, 2014.

532168 Optimization of Calcium Phosphate-Polymer Nanoparticle System for Co-Delivery of microRNA-21 Inhibitor and Doxorubicin

Abstract:

The development of nano-scaled carriers as a drug delivery platform has made a tremendous progress in the fight against cancer. Despite the progress, drug resistance remains one of the major challenges of nanotechnology-based cancer therapeutics. Due to gradually developed resistance of traditional anti-cancer drugs, new drugs that can target specific carcinogenic pathways are being sought based on molecular biology, such as small molecules, antibodies, RNAs, and DNAs. Among these, microRNA inhibitors (miRi) hold great promise in cancer therapeutics by down-regulating the proteins promoting multi-drug resistance. miRNA is a short non-coding RNA that can effectively inhibit protein translation or degrade specific mRNA transcripts. One unique feature of miRNA is that one miRNA can regulate multiple gene expressions. A key miRNA can act as a master switch for multiple pathways. An approach of a NP design carrying miRNA-21 inhibitor (miR-21i) and chemotherapeutic agent is expected to effectively treat the current challenges of drug resistance. However, the synthesis of a co-delivery system encapsulating miRNA inhibitor and chemotherapeutic agent is quite challenging, due to the opposite nature of the hydrophilicity of RNAs and the hydrophobicity of anti-cancer drugs.

In this study, a NP drug delivery system using biocompatible poly (l-lactic acid) (PLLA) was used to encapsulate doxorubicin (Dox). In order to increase the hydrophilicity, the NP system was coated with both 1,2-distearoyl-sn-glycero-3-phosphoethanolamine-N-[amino(polyethylene glycol) (DSPE-PEG) and dioleoylphosphatydic acid (DOPA), an anionic lipid. Negatively charged RNAs can condense with positively charged calcium ions to form a precipitated solid complex by the electrostatic interactions. Then an excess amount of the positive charges is neutralized by coating the complex with an anionic lipid DOPA. Thus the surface coating of the CaP/RNA complex with the hydrophilic head of the anionic lipid protects hydrophilic RNA from hydrophobic polymer, and the hydrophobic tail end allows the complex to be encapsulated along with Dox inside the hydrophobic polymeric NP structure. By delivering miR-21i and Dox simultaneously, the miR-21i upregulated tumor suppressor genes such as PTEN and PDCD4, thereby sensitizing the cells towards Dox. The optimization of the NP system with respect to particle size, polydispersity, drug loading as a function of different ratios of DOPA and DSPE-PEG is studied and will be presented. The optimum lipid to polymer ratio was found to be >50% (wt.) while a molar ratio of 1 to 9 for DOPA to DSPE-PEG gave the most stable particle with a size of ~75 nm. This NP system is tested against the two breast cancer cells of MCF7 and MDA-MB-231, and their cytotoxic efficacy would be presented. The results show that this combination therapy can improve the therapeutic index.

538234 Nanoallergens: A Liposomal Diagnostic Platform for Platinum-Based Drug Allergies

Abstract:

Platinum-based antineoplastic agents, also known as platins, are the most commonly used chemotherapeutic drugs for the treatment of various cancers. Despite their usefulness in treating many cancers, it has been reported that a higher population of cancer patients undergoing platinum based chemotherapy has suffered from drug hypersensitivity reactions (DHR) following the treatments. Specifically, frequent doses of oxaliplatin and carboplatin result in a high rate of IgE-mediated DHR, resulting in more severe and detrimental allergic reactions. Currently, there is a severe lack of predictive diagnostic tools mainly because platinum-based drugs reduce the number of basophils and plasma cells due to myelosuppression effects so that specific IgE level measuring and basophil testing become less useful. Therefore, it is important to have a predictive diagnostic for drug allergies to monitor the patient during treatment. Here, we propose a nanoparticle-based liposomal diagnostic platform which we call nanoallergen to accurately diagnose and predict oxaliplatin and carboplatin drug allergies. Nanoallergen is a synthetic liposomal allergen that multivalently presents metabolites of oxaliplatin and carboplatin to facilitate crosslinking of IgE-FcÎµRI complex, inducing degranulation responses in a similar manner of physiological allergic responses. In order to investigate the feasibility of our design for diagnosis of platinum drug allergies, we performed *in vitro* cellular degranulation assays on human FcÎµRI-expressing RBL-SX38 using oxaliplatin and carboplatin displaying nanoallergens with 7 clinical samples taken from platin allergic subjects. The results showed that significant degranulation responses were triggered by the oxaliplatin and carboplatin presenting nanoallergens at as low as 250 pM as compared to non-platinum drug allergic patient samples. Overall, the results suggest that nanoallergens lead to the high likelihood of success in diagnosis of platinum-based drug allergies by overcoming current diagnostic limitations.

536441 A Novel Approach for the Synthesis of Metallic Nanoparticles on Top of a Tellurium Nanowire Using a Green Synthesis Approach for Biomedical Applications

Abstract:

Statement of Purpose: In a world of constant duality, where the population density increases as much drama as the natural resources disappear, there is some concern that should be addressed from the chemical point of view. Traditional chemistry has undoubtedly enhanced the quality of life, but a price of the threaten of the environment and the health of humans. Chemical processes are responsible for the contamination of all the known ecosystems all over the planet, to more or less extent. The traditional physics-chemical approaches for the synthesis of compounds and drugs that we used for concerns such as the antimicrobial resistance of the fight against cancer are easy-to-get straightforward. Nevertheless, there is a cost associated with the limitations that should be overcome from these approaches, such as the production of toxic by-products or the biocompatibility of the products. Therefore, new methods are needed, and the green chemistry offers itself as a suitable and novel answer, achieving a safe and environmentally-friendly design, manufacture and use of chemical products.

Methods: Tellurium nanowires were prepared using both chemical and green chemistry approaches. For the first one, sodium tellurite was dissolved in an aqueous solution of polyvinylpyrrolidone (PVP), hydrazine monohydrate and ammonia, and keep inside a hydrothermal reactor for several hours. The environmentally-friendly approach led to the use of telluric acid and starch as a unique reducing agent and kept inside a hydrothermal reactor for several hours. Both products were purified using the same conditions and characterized regarding composition, structure, and morphology. Once purified, tellurium nanowires were used as a template for the growth of metallic nanoparticles (such as silver, platinum, and palladium) in a quick method with no need of additional reducing agent at room temperature. Besides, biocompatibility tests of the tellurium nanowires with human tissue were accomplished, growing human dermal fibroblast (HDF) cells in media in the presence of both chemical and green-synthesized nanowires. After an incubation time of 5 days, the cell growth was analyzed using MTS assay. Furthermore, anticancer studies were done using the tellurium nanowires with melanoma cells for 5 days.

Results: It is demonstrated that green synthesized tellurium nanowires can be used as a template for the growth of spherical metallic nanoparticles in a quick reaction that takes places in 1 minute, at room temperature with no need of stirring and reducing agent. TEM images of both tellurium nanowires and metallic nanoparticles were taken, showing nanoparticles with a constant distribution size attached to the tellurium nanowires, which were quickly released from the structure. The chemistry of the samples was confirmed using EDX analysis, showing the distinct peaks of tellurium and the ones corresponding to each one of the metals. In vitro cytotoxicity assays were performed with human dermal fibroblasts (HDF) cells. The experiments showed that green-synthesized nanowires showed a higher cell proliferation of the cells than the experiments carried out with chemically-synthesized structures. Furthermore, after the incubation period, the growth of the cells was enhanced in comparison with the control and the chemically-synthesized nanowires. Anticancer studies showed an enhanced anticancer activity of green nanowires compared to the chemically-synthesized structures over a period of 5 days, causing a higher depletion of cell viability.

Conclusions: Current and main methods to synthesize nanostructures, both nanowires and nanoparticles, use approaches which employ chemical methods. The weakness of these procedures (extreme reaction conditions, production of toxic byproducts, and need of purification) calls for a necessity of alternative approaches. Green chemistry can be used to overcome these drawbacks. Here, green-synthesized tellurium nanowires were compared with chemically-synthesized structures to show that the first ones have an enhanced biocompatibility and anticancer properties over the ones synthesized using traditional methods. Besides, it was demonstrated that tellurium nanowires synthesized using starch can be used for the controllable and quick growth of metallic nanoparticles.

532417 Formulation and Recovery of Fast-Acting Lumefantrine Nanoparticles for Oral Malaria Therapy

Abstract:

Malaria is a major source of mortality in developing tropical regions and new low-cost therapies are needed. Lumefantrine is a Biopharmaceutics Classification System class II drug, and increasing the bioavailability of lumefantrine has the potential to reduce the dose and number of required administrations per treatment, thus reducing the financial burden of malaria therapy. Formulating therapeutics into nanoparticles can increase their bioavailability by reducing drug crystallinity and increasing surface area, thereby resulting in faster dissolution kinetics. In this study, we use Flash Nanoprecipitation to encapsulate lumefantrine into nanoparticles with low-cost stabilizers (zein/casein, HPMCAS, and lecithin) and assess the release of lumefantrine in simulated gastric and intestinal fluids. For the low-melting-point lumefantrine, a feasible spray-drying protocol is further developed to dry lumefantrine-encapsulated nanoparticles into powders, while maintaining the physical state of the drugs. The release experiments show that lumefantrine formulations with all stabilizers showed superior dissolution compared to crystalline lumefantrine release in simulated fasted and fed intestinal media. Remarkably, particles synthesized with zein/casein and HPMC exhibited near complete release in under one hour. These cheap and efficient formulations have the potential to be developed into future therapies for malaria.

531990 Engineering Antibacterial Nanosurfaces for Field Hospitals

Abstract:

<>Antibiotic resistance (ABR), although not a new phenomenon, is rapidly gaining attention within the scientific community as an escalating threat to the global health care infrastructure. The increasing frequency of occurrence of resistant bacterial strains, specifically multidrug resistant (MDR), extensively drug resistant (XDR), and totally drug resistant (TDR) phenotypes, would seemingly necessitate demand for a new generation of small-molecule drugs for clinical implementation. Unfortunately; the compound effect of administrative barriers, technical complications, and selective economic pressures has significantly reduced the rate of discovery for new antibiotic species. In addition to this, systemic overuse and misuse of currently available therapeutics is generally accepted as the principle driving force behind the induction of resistance. This is a crucial factor that has yet to be effectively mitigated through international regulation, and would only be perpetuated with the development of new classes of antibiotics. While attempting to develop new engineering solutions that surpass the specified shortcomings, novel approaches towards the management of bacterial growth and colonization in clinical settings are highly desirable.

One such approach is the development of nanoscale surface functionalization for materials which are commonly seen in a clinical setting, ranging from general architectural elements to instrumentation. Nanoscale functionalization, which includes both physical and chemical components interacting with biological systems, in this work has been motivated by macrophage biomimetics. When macrophages phagocytize bacteria for systemic clearance, the bacteria are entrained within endosomal compartments which subsequently mature with a reduction in pH, concentration of enzymatic species, and generation of reactive oxygen species (ROS). Although these factors work synergistically to induce cytotoxicity within the complex environment of the endosome, ROS generation along nanoparticulate surfaces (in addition to other nanoscale phenomena present within colloidal systems) has prove sufficient in demonstrating antibacterial properties *in vitro* and *in vivo*. Although metallic nanoparticulates, specifically silver, are frequently implemented within antibacterial applications, this work opted to utilize selenium nanoparticles for achieved therapeutic effect. Selenium was selected on the grounds that its nanoparticulate formulations are capable of ROS generation within cellular targets, while still serving as an overall trace elemental nutrient in the human diet.

Selenium nanoparticle functionalization was achieved through heterogeneous nucleation along a diverse range of solid supports provided by academic and federal collaborators. These materials were coated in small-batch processes, in which the solid supports were immersed in a reaction solution for 60 seconds, followed several successive washing stages with deionized water. Nanoparticulate coatings were characterized with atomic force microscopy (AFM), scanning electron microscopy (SEM), energy dispersive X-ray spectroscopy (EDS), X-ray photoelectron spectroscopy (XPS), and contact angle measurement. Physical characterization revealed that functionalization yielded heterogeneous surface coverage of nanoparticles, with average interspacings within the sub-micron range. Chemical characterization verified that the observed nanoparticulates were composed of elemental selenium. Following physical and chemical characterization, functionalized surfaces were assessed for antibacterial properties by means of colony forming unit (CFU) assay. Treated materials were immersed in cultures of gram-positive (*Staphylococcus aureus*) or gram-negative (*Escherichia coli*) bacterial strains, and the concentration of viable, adsorbed bacteria were determined at set intervals following initial exposure. Statistically significant reduction in *S. aureus* bacterial colonization was observed for the two current test materials, carbon-microfiber sheets and basalt-epoxy blocks. For carbon-microfiber sheets, nanoparticulate treated surfaces resulted in 32 % reduction and 63 % reduction in CFU at the material's targeted operational temperatures of 50 degrees Celsius and 40 degrees Celsius, respectively. For basalt-epoxy blocks, nanoparticulate treated surfaces resulted in a 77 % reduction and 82 % reduction in CFU at the material's targeted operational temperatures of 60 degrees Celsius and 50 degrees Celsius, respectively. A general trend of reduced bacterial colonization has been observed for *E. coli* in preliminary studies.

The future directions of this project are focused on elucidating the mechanism of antibacterial activity, assessing mammalian cell biocompatibility, determining efficacy under biological fouling, and investigating the general mechanical properties of nanoparticle functionalized composites. The mechanism of antibacterial activity is of significance, as this may pose serious implications on the potential for subsequent bacterial resistance. While the understanding of resistance developed under the selective pressures of antibiotics is historically well established, current studies in the literature have demonstrated select cases for the development of resistant phenotypes in bacterial populations exposed to silver nanoparticle treatments. It is of great interest to determine if this pattern is similarly observed with the implementation of selenium nanoparticulate coatings. As ROS are believed to serve as a principle factor in prokaryotic cytotoxicity, which is inherently non-specific for select biomolecular targets within pathogens, there is additional interest in determining whether selenium nanoparticles are either cytotoxic or mutagenic to various eukaryotic cell lines. In addition to this, if ROS generation is crucial for achieving antibacterial efficacy, than direct exposure of colonizing bacterial cells to the nanoparticulate coating is essential. Steric constraints on this exposure may result from the formation of a fouling layer, arising either through the directed-assembly of bacterial biofilms or the random deposition of organic/inorganic debris, and thus there is also interest in determining the antibacterial properties of selenium nanoparticulate coatings under biofouling conditions. Finally, the mechanical properties of nanoparticulate coated composites will also be assessed under variant environmental conditions, such as extreme heat and humidity, that may be experienced in field hospitals operating in diverse locations throughout the globe.

Figure One: Appearance and antibacterial property of basalt-epoxy blocks with selenium nanoparticle coating. (A.) An SEM image of a basalt-epoxy surface following selenium nanoparticulate functionalization. Observe that the nanoparticles that have nucleated along the surface are polydisperse, both in their sizing and morphology. The interspacings between nanoparticulates are smaller than the dimensions of common bacteria, such as *S. aureus* and *E. coli*. Nanoparticulates have even formed within crevices along the treated surface, demonstrating that coverage extends to all areas that are accessible to the reactant solution during particle nucleation. The inlaid scale bar is 2 micrometers in length. (B.) The antibacterial efficacy of selenium nanoparticulate coatings of basalt-epoxy blocks at two targeted operational temperatures set by collaborators. Functionalization with selenium nanoparticles yielded a statistically significant reduction in bacterial colonization at both temperatures, as demonstrated by CFU assay. Treated surfaces experienced a 77 % and 88 % reduction in CFU at the targeted temperatures of 60 degrees Celsius and 50 degrees Celsius, respectively. N = 3.

535117 Overcome Drug Resistance of Cancer Cells By Confining, Perturbing and Analyzing Them in Nano-Liter Chambers One Cell at a Time

Abstract:

Tumor cells develop resistance to many therapies. And it is a major problem across many non-surgical cancers. We find that, multi-functional nanotechnology, when applied onto drug resistance development scenario, could uncover the resistance mechanism and provide rationale for designing effective combination therapy to overcome it.

Although it may take a few months to see the reoccurrence of tumor, yet people recently realize that tumor cells already developed resistance as early as a few days after drug treatment. This is because that tumor cells can easily change its cell state from the original drug sensitive one to a very different one, which allow them to survive under the drug treatment. Therefore, stop the unfavorable cell state transition is required in order to overcome drug resistance. To do that, it is essential to discover the driver regulators that lead the unfavorable cell state changes. Finding the driver regulators are very hard because of 2 major obstacles. Because there are mixtures of different cell states involved in such transition (need to analyze individual cells), and because, within each cell, the regulatory network has many regulators involved (need to characterize many proteins), single-cell proteomic analysis is the best way to study it.

However, single-cell proteomics are extremely hard to implement mainly because the protein amounts in single cells are too little to be detected. Fortunately, this problem could be beautifully addressed by a multi-functional bionanotechnology developed in our lab called single-cell barcode chip (SCBC). This nanotechnology can confine individual cells in nano-liter chamber. Perturbing and lysing single cell within the chamber is also feasible through programming the nano-liter fluid. Consider the nano-liter size of each chamber, small amount of protein from single cell when released within such tiny volume will also have a detectable concentration. Based on such multi-functional, programmable, nano-chamber, many important regulators from each individual cell are quantified simultaneously with the help of arrays of antibodies and nano-probes installed orderly within each chamber.

When applied to tumor cells that are under therapeutic treatment, this powerful nanotechnology enables us to uncover the changes of many important regulators simultaneously across thousands of individual cancer cells. The big data generated from the nanotechnology pinpoint the most important regulators that are driving the initiation of such unfavorable cell state changes, which leads to resistance. Base on the information from the powerful nanotechnology, an effective combination therapy is designed which is experimentally validated to be able to halt the unfavorable cell-state transition and therefore arrested the resistance development.

This powerful nanotechnology could provide deep biological insights that are otherwise impossible to see. The nanotech has the potential to uncover the resistance mechanism across many other cancer types and potentially bring durable therapy response to a broad range of cancer patients. This may usher a new era of nanotech-assisted rational designing of combination therapy.

534690 Optimization of Liposome-Hollow Gold Nanoparticle for mRNA Delivery

Abstract:

mRNA therapy has gathered great interest because of its potential role for the transient modification of immune cells for treatment of an array of diseases. However, the main obstacle preventing routine mRNA immune cell transfection is the lack of a high viability, high efficiency, high throughput delivery system. Due to its sensitivity towards catalytic hydrolysis, unprotected mRNA cannot be delivered by itself. We propose encapsulating mRNA in neutral liposomes decorated with cell internalizing peptides and hollow gold nanoshells (HGN). We use passive entrapment of lightly condensed mRNA in phospholipid-cholesterol bilayer liposomes incorporating polyethylene-glycol-lipid tethered TAT and HGN. Following TAT induced endocytosis, we can irradiate the cells with picosecond pulsed near infra-red light to induce vapor nanobubbles around the HGN that can rupture both the liposomes and the endosomes, thereby releasing the MRNA to the cell cytoplasm, where the mRNA can produce the proteins of interest. We will present our work toward the realization of this light triggered mRNA delivery system.

522523 Lipid Nanoparticle Ionization at Endosomal pH Is a Cell-Free Predictor of mRNA Delivery Efficacy In Vivo

Abstract:

Messenger RNA (mRNA) therapy has the potential to transform the lives of millions of people suffering from a wide range of diseases by augmenting protein expression or replacing missing or mutated proteins. The versatility of mRNA-based drugs has made it a leading gene therapy platform, with dozens of studies utilizing it for applications such as such as protein replacement, prophylactic vaccines, cancer immunotherapy, gene editing, and cellular reprogramming. However, the clinical translation of mRNA therapeutics is contingent upon their safe and effective delivery into target cells. Lipid nanoparticles (LNPs) have emerged as a promising class of delivery materials with the potential to facilitate mRNA delivery. Given the high cost of preclinical testing, most delivery systems are screened in cell culture before advancing into murine and non-human primate models. Unfortunately, cell culture testing is resource intensive and prone to high rates of false negatives and positives. Here, we describe an alternative, cell-free screening strategy that accurately predicts mRNA delivery potency in animals. This was accomplished through the synthesis of a small library of lipid-like (lipidoid) materials, a subset of which effectively delivered mRNA to the liver and spleen of mice following systemic injection of mRNA-LNPs at a 0.75 mg/kg dose. In particular, the lipidoid $306O_{i10}$, which contains a branched tail, facilitated marked levels of firefly luciferase expression in mice. A structure-function analysis using our *in vivo* results revealed that *in vitro* efficacy, particle size, mRNA entrapment, and pKa did not correlate with efficacy in mice. However, the degree of lipid nanoparticle surface ionization at the late endosomal pH of 5.0 was a strong predictor of *in vivo* efficacy (R^2= 0.95). The generation and testing of a secondary library of 33 lipidoids confirmed LNP surface ionization as an effective screening tool, correctly predicting materials with both high and low efficacy in mice. These data indicate that the measurement of nanoparticle surface ionization is an inexpensive, high throughput, and reliable method to identify mRNA delivery materials that are highly potent *in vivo*.

536256 pH Responsive Polycationic Nanoparticles for siRNA Delivery in Inflammatory Bowel Diseases

Abstract:

Title: pH responsive polycationic nanoparticles for siRNA delivery in Inflammatory Bowel Diseases

Authors: Aaliyah B. Shodeinde[1,2], Angela M. Wagner[1,2], Noor Al-Sayyad[3] and Nicholas A. Peppas[1,2,4,5,6]

[1] McKetta Department of Chemical Engineering, The University of Texas at Austin, Austin, TX, USA

[2] Institute for Biomaterials, Drug Delivery, and Regenerative Medicine, The University of Texas at Austin, Austin, TX, USA

3 Department of Physics, The University of Texas at Austin, Austin, Texas, USA

[4] Department of Biomedical Engineering, The University of Texas at Austin, Austin, Texas, USA

[5] Department of Pediatrics, and Department of Surgery and Perioperative Care, Dell Medical School, The University of Texas at Austin, Austin, TX, USA

[6] Division of Molecular Pharmaceutics and Drug Delivery, College of Pharmacy, The University of Texas at Austin, Austin, TX, USA

The RNA interference (RNAi) mechanism has become an increasingly desirable approach to elicit gene knockdown. Specifically, small interfering RNAs (siRNA) have been successfully incorporated within the native RNAi machinery to induce the degradation of undesirable mRNA in multiple pathogeneses. Due to the ease of enzymatic degradation of siRNA in addition to their propensity for renal clearance upon systemic administration, it is essential to develop novel carriers to ensure the delivery of the payload to the desired site of action. We report a novel method for the synthesis of a methacrylate based polycationic nanoparticle system in aqueous media. These nanoparticles have been shown to exhibit lower pKa values than previous cationic nanoparticle formulations, preventing premature swelling upon interaction with the intestinal mucosa. Furthermore, greater localization of the siRNA payload within the nanoparticle core has been observed which ensures siRNA protection pending cellular internalization.

The physical properties of the resulting nanoparticles were characterized using dynamic light scattering, zeta potential, titration, pyrene fluorescence, and red blood cell hemolysis to deduce the influence of polymer composition on surface charge, pKa, swelling ratio, hydrophile-hydrophobe phase transition, and erythrocyte membrane disruption capability. In vitro biocompatibility was assessed with relevant cell lines using MTS cell proliferation and LDH membrane integrity assays. The delivery capacity of the nanoparticles was analyzed using multiple agents for loading and release studies.

537761 Lipid Nanoparticle Mediated Drug Delivery for Targeting Inflammation Site in Atherosclerosis

Abstract:

Atherosclerosis is an arterial disease that involves build-up of plaques composed of cholesterol and macrophages inside the walls. This process is associated with attachment of the low density lipoprotein (LDL) particles to the proteoglycan of endothelium at the site of infection. These LDL particles are highly susceptible to oxidation and further induce inflammation by increasing the leukocyte adhesiveness and permeability to the endothelium. This condition may lead to complications like myocardial infarction or cerebrovascular accident which are leading causes of death in the world. Targeted drug delivery using nanoparticles has several advantages over the traditional drug delivery method such as reduced dosage of drug, lesser side effects, decrease in cost of therapy and increased accumulation of drug within the target tissue. Liposomes are very biocompatible and highly tunable for specific applications. Thus they can be loaded with suitable drugs and can be efficiently used for targeting the inflammation site in atherosclerosis. In this project, we are developing a highly customizable lipid based nanoparticle drug delivery vehicle for targeting inflammation sites of atherosclerosis plaques. We have engineered lipid nanoparticles (LNPs) of various sizes, surface charges, and coating density of hydrophilic polymer using non-cationic lipid and cholesterol components. LNPs were characterized by both photon correlation spectroscopy to determine particle size distribution, hydrodynamic diameter, and polydispersity, and by zeta potential measurements to determine nanoparticle surface charge characteristics and to quantify colloidal stability. Our preliminary cell uptake studies using FITC tagged LNPs indicate the nanocarriers were internalized into the cytosol and provided confidence that we target endothelial and smooth muscle cells in the atherosclerotic environment. A comparative study is underway to determine the targeting efficiency of antibody tagged LNPs to endothelial and smooth muscle cells. We are investigating cellular response with optical microscopy, and probing biochemical changes using a novel 3D co-culture platform of endothelial cells and smooth muscle cells we have developed. Further, we will be testing Atorvastatin drug encapsulated particles in combating oxidative stress in endothelial and smooth muscle cells. The inhibitory effects on production of the proinflammatory cytokines were studied following the nanoparticle targeting experiments. A significant decrease in production of proinflammatory cytokines was observed when the cells were treated with drug loaded lipid nanoparticles. The future work of this research involves optimization of the process along with ex-vivo studies using the inflamed artery extracted from pig.

535729 Internalization and Endocytic Trafficking of 3WJ RNA Nanoparticles for siRNA Delivery

Abstract:

RNA nanotechnology is becoming increasingly successful as a therapeutic agent for the treatment of cancer and genetic diseases. The intrinsic features of RNA allow the bottom-up assembly of complex structures and nano-scale devices in numerous geometries. Though the RNA structure is conceptually similar to DNA, it offers elasticity, flexibility, and the ability to form non-canonical base pairing. Guo and co-workers have constructed RNA nanoparticles with controllable shape, size, and stoichiometry using the stable phi29 packaging RNA three-way junction (3WJ) motif as a scaffold, which displays high chemical and thermodynamic stability. A benefit of these particles is the ability to functionalize them with targeting agents, therapeutic siRNA/miRNA, and fluorescent imaging molecules. These particles have shown success in preliminary studies *in vivo* but the mechanisms of cellular entry and intracellular trafficking have not been explored. This project aims to elucidate these mechanisms in order to inform rational design to improve particle targeting specificity and survive endosomal processing.

RNA cellular internalization through endocytosis occurs primarily through clathrin-dependent, caveolin-dependent, and macropinocytosis pathways. These pathways can be inhibited through pharmacological small-molecule inhibitors, which can give insight into the impact of each pathway on particle internalization and subsequent endosomal processing. A variety of endosomal assays can be used to track the processing of particles. However, this report is interested in whether or not the particle uptake pathway leads to an acidified endosome. Bafilomycin A1 (BafA1) inhibits endosomal proton ATPases, which prevents the acidification of the endosome. Thus, transfections in the presence of BafA1 can elucidate the reliance of particle survival on the endosomal environment, and fluorescent co-localization reveals visual proof of particle sequestration to pathway-specific endosomes.

3WJ RNA nanoparticles comprise a siRNA targeting the luciferase reporter gene, an Alexa647 fluorescent probe, and an epidermal growth factor receptor-targeting aptamer (EGFR). Particles are transfected into MDA-MD-231 cell line stably expressing the luciferase reporter gene (MDA-luc). Protein knockdown is measured by a decrease in bioluminescence when lysate is exposed to the luciferase substrate. The dose providing highest knockdown is used in transfections preceded by endocytic pathway inhibition by small molecule inhibitors to elucidate pathway impact on particle efficiency. Similarly, flow cytometry is used to determine particle uptake after pathway inhibition. BafA1 is then used to determine the impact of acidic environments on the uptake and knockdown efficiency of particles entering through various pathways.

The EGFR receptor is known to internalize through clathrin-dependent endocytosis. Therefore, inhibition of the clathrin-dependent endocytic pathway is expected to result in a compensatory uptake through alternative pathways. Previous reports show that, when clathrin is inhibited, delivery is enhanced. This study is expected to follow this trend. Inhibition through BafA1 is expected to have a positive effect on knockdown when particles are processed by acidified endosomes (clathrin-dependent endocytosis and macropinocytosis) and no effect when not (caveolin-dependent endocytosis). These implications suggest that increased delivery and knockdown efficiency can be achieved through the targeting of alternative, non-acidified pathways or appending endosomal buffering moieties to the particles that inhibit acidification.

535648 Synthesis of Poly(Aspartic Acid)-Doxorubicin Prodrug for Sequential Delivery of Afatinib and Doxorubicin

Abstract:

Combination therapy has a great potential for enhancing the efficacy of chemo drugs in treating cancer. Drug combinations that exploit the blockage of multiple survival pathways in cancerous cells hold great promise in achieving synergistic effects. Recent studies indicate that cancer cells can remarkably be sensitized to the effects of DNA-damaging drug, doxorubicin (Dox), after epidermal growth factor receptor (EGFR) signaling is suppressed. To achieve the maximum efficiency while minimizing the fast clearance and systemic toxicity, EGFR inhibitor needs to be released first while Dox is prevented from premature leakage via conjugation to a polymer to form a polymeric prodrug. In this study, we synthesized two poly(aspartic acid)-Dox prodrugs using different linkages: a stable amide bond (PAsp-Dox) and a pH-sensitive hydrazone bond (PAsp-pH-Dox). PAsp is an amino acid with a carboxylic group available at each side chain for the conjugation of many Dox molecules. Both PAsp-Dox and PAsp-pH-Dox prodrugs can be easily encapsulated in biocompatible and biodegradable poly(l-lactic acid)-polyethylene glycol (PLA-PEG) along with afatinib (Ai), which is complexed with dioleoyl phosphatidic acid (DOPA) via ion paring. By physically blending the two PAsp-Dox and PAsp-pH-Dox prodrugs at different ratios, we could get the desired programmed drug release profile. Both Dox and Ai were encapsulated inside the PLA-PEG with almost full encapsulation efficiency (EE >~90%) and high drug loadings (>~8%(wt)). The size of nanoparticles (NPs) encapsulating various different combination drugs ranges between ~70-80 nm, measured by dynamic light scattering (DLS) and transmission electron microscopy (TEM). Our NP design has shown great promise for the sequential co-delivery with sustained release through results for release kinetics in phosphate-buffered saline (PBS) in pH 7.4 and 5 and *In vitro* cytotoxicity against three non-small cell lung carcinoma (NSCLC) cell lines, NCI-H358, NCI-H1650, and A549. This NP system can improve the therapeutic index by achieving sequential delivery with a desired time delay and reducing systemic exposure of drugs through decelerated blood clearance and high tumor accumulation.

536672 Co-Delivery of 2-DG and V9302 Via a Prodrug Micellar Formulation for Synergistic Targeting of Metabolism in Cancers

Abstract:

The unique metabolic demand of cancer cells suggests a new therapeutic strategy targeting the metabolism in cancers. V9302 is a recently reported ASCT2 amino acid transporter inhibitor which shows promising antitumor activity by blocking glutamine uptake. However, its poor solubility in aqueous solutions and tumor cellsâ€™ compensatory metabolic shift to glucose metabolism may limit the antitumor efficacy of V9302. 2-Deoxyglucose (2-DG), a derivative of glucose, has been developed as a potential antitumor agent through inhibiting glycolysis in tumor cells. In order to achieve enhanced antitumor effect by inhibiting both metabolic pathways, a 2DG prodrug-based micellar carrier (POEG-p2DG) was developed. POEG-p2DG well retained the pharmacological activity of 2-DG in vitro and in vivo. More importantly, POEG-p2DG self-assembled to form micelles that were capable of loading V9302 to achieve co-delivery of 2-DG and V9302. V9302-loaded POEG-p2DG micelles were small in sizes (~10nm) and showed a slow kinetics of drug release. In addition, V9302-loaded POEG-p2DG micelles exhibited improved cytotoxicity towards tumor cells compared to free V9302. These results suggest that POEG-p2DG prodrug micelles may serve as a dual functional carrier for V9302 to achieve synergistic targeting of metabolism in cancers.

537740 Engineering PEO-Pdlla Nanoparticles Containing the PI3K Inhibitor LY294002

Abstract:

The PI3K pathway is critical for many cellular processes from cell cycle regulation and apoptosis to DNA repair. The PI3K pathway also happens to be one of the most consistently dysregulated signaling pathways in cancer cells (1). Targeting this pathway has the potential to create more effective therapies for a variety of neoplastic diseases. LY294002 is a PI3K inhibitor that has shown promise in *in vitro* models but has not successfully transitioned to *in vivo* models due to its poor water solubility. We hypothesized that encapsulating LY294002 into polymer nanoparticle using the Flash Nanoprecipitation (FNP) technique would lead to high drug loadings and overcome the current barrier preventing *in vivo* studies.

Poly(ethylene oxide)-b-poly(D,L lactic acid) (PEO-PDLLA) nanoparticles containing LY294002 loaded at > 10 wt% were fabricated using FNP with sizes ranging from ~ 150 â€“ 250 nm and zeta potentials > -20 mV in DI water. The PEO and PDLLA molecular weights in the diblock copolymer result in PEO brush layers at the particlesâ€™ surface and steric stabilization. This drug delivery system has sufficient LY294002 loading to provide clinically relevant LY294002 dosing and allows for the translation of LY294002 into the *in vivo* setting. This presentation will focus on the design, synthesis, and characterization of these nanoparticles. The flexible particle fabrication methods described here can be applied to a wide range of therapeutic agents, and the use of FNP allows for future scale-up of nanoparticle production.

1. Fruman DA, Rommel C. PI3K and cancer: lessons, challenges and opportunities. Nat Rev Drug Discov. 2014;13(2):140-56. doi: 10.1038/nrd4204. PubMed PMID: WOS:000331145900018.

536333 Targeting Tumor Associated Macrophages with PAMAM Dendrimers Improves Therapeutic Efficacy in Glioblastoma

Abstract:

Introduction: Cancer is the second leading cause of death in the United States, and brain cancers are among the most severe types of cancers with poor patient outcomes due to barriers associated with delivering therapeutics to the brain in addition to challenges associated with delivering drugs to solid tumors.[1] Many anti-cancer therapies fail to fully penetrate tumor tissue and access tumor cells at relevant concentrations due to ill-developed vasculature and high interstitial pressure within the tumor.[2] Additionally, the blood brain barrier (BBB), which under healthy regimes is a natural defense mechanism stopping foreign substances from reaching the brain, prevents therapeutically relevant quantities of treatments from reaching the brain tumor.[3] As a result, patients with central nervous system cancers have failed to see improved outcomes despite the discovery of powerful new anti-cancer drugs because these therapies 1) have poor BBB penetration and 2) fail to distribute and accumulate within the tumor. This two-pronged barrier necessitates the administration of high doses of systemic therapies to achieve effective doses in the brain tumor, which cause significant toxicity in the body. Therefore, a vehicle that carries therapies to the brain, into the brain tumor, and to the tumor associated macrophages (TAMs) while remaining inactive in the body provides a promising clinical strategy.

Methods: Our group has identified and extensively characterized hydroxyl polyamidoamine (PAMAM) dendrimers as promising nano-vehicles for targeted drug delivery in several small and large models of neurodegenerative disorders.[4,5] Here, we present these dendrimers for targeting brain tumors in a mouse model of glioblastoma (GBM). C57BL/6 mice 6-8 weeks old were inoculated with 100,000 GL261 murine GBM cells by intracranial injection. To assess tumor targeting, generation 4 PAMAM dendrimers were administered systemically 14 days post-inoculation and analyzed via confocal microscopy and fluorescence spectrometry. In addition, two therapies were evaluated: BLZ945, an immune-modulating agent, and triptolide, an anti-proliferative drug. These therapies were conjugated to dendrimers via pH-sensitive linkers for triggered release. With BLZ945, mice were administered with a single systemic dose of free or dendrimer-conjugated drug 10 days post-inoculation at 100 mg/kg on a drug basis. With triptolide, mice were systemically administered daily at 0.5 mg/kg free or dendrimer-conjugated drug starting 5 days post-inoculation. Mice were monitored daily and their disease progression assessed with kyphosis scoring and in open field trials.

Results: We demonstrate that upon systemic administration, PAMAM dendrimers (without any targeting ligands) are able to cross the BBB, diffuse freely through the brain parenchyma, and distribute uniformly throughout the brain tumor. Confocal imaging shows that dendrimers selectively localize to Iba1-labelled (TAMs) while exhibiting minimal uptake by resting microglia/macrophages in healthy brain tissue (**Fig.1**). Quantitatively, dendrimers accumulated specifically within the tumor at >5-fold greater levels compared to the contralateral hemisphere.

We next leveraged this tumor targeting capacity to deliver therapeutics for enhanced efficacy, particularly in the context of favorably manipulating TAMs. We find that a single systemic dose of dendrimer-BLZ945 immunotherapy halfway through disease progression prolongs survival by >25% compared to BLZ945 and control treatment groups. Additionally, dendrimer-BLZ945 treatment delayed the onset and severity of kyphosis, as well as improved motor function in open field tests. This effect is mediated by a reduction in expression of Arginase-1, a marker of anti-inflammatory tumor promoting phenotype, indicating an enhancement in immune-modulating effect with dendrimer delivery. For triptolide, targeted delivery with the dendrimer resulted in significantly reduced tumor size compared to free triptolide, as well as improvements to motor function and kyphosis.

Conclusions: PAMAM dendrimers (without any targeting ligands) selectively target TAMs within the GBM tumor upon systemic administration and show minimal accumulation in healthy brain tissue. This targeting capacity results in improved efficacy of immunotherapy BLZ945 and anti-proliferative triptolide compared to their free drug counterparts. We are currently further assessing and optimizing these dendrimer-mediated therapies, with the plan for combination therapies in the future. Therefore, PAMAM dendrimers have significant clinical applications for the treatment of GBM by selectively delivering therapies to the tumor, particularly in the context of localized manipulation of the immune profile in GBM tumors.

References:

[1]Siegel, R.L., Miller, K.D, Jemal, A. *CA: A Cancer Journal for Clinicians*. 2018. 68(1):7-30

[2]Tredan, O., Galmarini, C.M., et al. *J of the Nat Cancer Inst*. 2007. 99(19): 1441-54

[3]van Tellingen, O., Yetkin-Arik, B., et al. *Drug Resistance Updates*. 2015. 19: 1-12

[4]Kannan, S., Dai, H., Navath, R.S., et al. *Sci Trans Med*. 2012. 4(130): 130ra46

[5]Mishra, M.K., Beaty, C.A., et al. *ACS Nano*. 2014. 8(3): 2134-47

532365 Programming Tumor-Clearing Macrophages with Targeted Gene Therapy

Abstract:

Macrophages (mφs) are key immune effectors that infiltrate into tumor in high numbers. However, within the immunosuppressive tumor milieu, they undergo a switch from an activated (M1) tumoricidal state to an immunosuppressive (M2) phenotype, which facilitates tumor growth and metastasis. Much effort has therefore been devoted to developing therapies that target tumor-associated mφs (TAMs) as an alternative to classic radio/chemotherapy. *To date, only systemic cytokine blockade using antibodies or small molecule drugs has shown success in this arena;* however, because these blockades have low response rate among patients and suppress all mφs in the body, they also induce dangerous side effects.

Instead of ablating mφs through cytokine inhibition, we genetically reprogramed suppressive M2 mφs in situ into highly effective, tumor-clearing M1 mφs by delivering genes encoding transcription factors. Specifically, we used mannose receptor-targeted nanoparticles (NPs) to provide mφs with mRNAs encoding master regulators of mφs polarization - Interferon Regulatory Factor 5 (IRF5). To identify the critical changes in mφs gene expression associated with NPs treatment, we used genome-sequencing technology to analyze 770 genes in 19 different pathways and processes in myeloid cells. Our results demonstrated that when NPs carrying mRNA encoding IRF5 (IRF5-NPs) were used treat interleukin 4 (IL-4) exposed 'M2-like mφs', it promoted signature genes associated with M1 mφs while suppressed signature genes associated with M2 mφs. These signature genes persisted even when mφs were re-challenged with IL-4 media.

In an ID8 cells-induced syngeneic mouse model of ovarian cancer, repeated intraperitoneal (i.p.) injection of IRF5-NPs significantly suppressed tumor growth in all tested mice and demonstrated survival benefits in 40% tested mice. To determine the mechanism of IRF5-NPs suppressing tumor growth in this model, we analyzed the immune cells population in mouse peritoneum through flow-cytometry. IRF5-NPs treatment suppressed the expansion of M2 mφs population in the peritoneum of ovarian tumor mice, while maintaining a constant M1 mφs population. In a more aggressive B16F10 lung metastasis mouse model, repeated intravenous (i.v.) injection IRF5-NPs delayed tumor growth by 50%. Based on histology analysis, complete blood count and serum chemistry analysis, we showed that repeated injection of NPs under therapeutic doses did not induce serious toxicity to healthy mice.

This research is a pioneering effort to treat cancer by genetically reprogramming suppressive M2 mφs. The results of the project could have a transformative effect on cancer therapeutics by **(1)** providing a foundation for gene-modification systems that would enable physicians to obviate the suppressive tumor milieu. **(2)** It also has the potential to work synergistically with existing immunotherapies by creating a therapeutic window for cancer patients, thus stimulating a stronger overall immune response in patients who also receive immunotherapy. **(3)** Although our project focuses on mφs, genetic re-engineering using NPs could also be applied to reprogram other tumor-promoting cell types such as regulatory T cells, MDSCs. Thus, the approach could provide a basis for future initiatives involving collaborations at the

interface of materials science, gene therapy, and immunology.

525015 Atomically-Precise Van Der Waals Heterostructures of Graphene and h-BN for 2D Circuits

Abstract:

Van der Waals heterostructures of isostructural and isoelectronic graphene and hexagonal boron nitride (h-BN) offer a singular platform to study fundamental physics phenomena such as the origin of energy bandgap opening at the Dirac point in graphene, Hofstadterâ€™s butterfly effect in graphene, chiral quantum states of Dirac electrons, and the detection of topological currents. Such phenomena in graphene/h-BN or graphene/h-BN/graphene or h-BN/graphene/h-BN planar sp^2 lattice stacks are extremely sensitive to their process of synthesis and nanoscale assembly. Current techniques mostly rely on the mechanical or chemical transfer of both h-BN and graphene to build the 2D sandwich heterostructures, which introduces the interfacial polymeric and metallic heteroatom contaminations. Here we report the direct, transfer-free, and atomically-precise development of both vertical and lateral heterostructures of graphene and h-BN *via* a bottom-up all-chemical vapor deposition (CVD) strategy. This 3-step approach combines: (i) the catalyst-free nucleation of h-BN on oxide and nitride-based gate-dielectric substrates *via* surface chemical-interaction guided mechanism; (ii) physical deposition of a thin, polycrystalline, and low-carbon solubility metal film on h-BN; and (iii) synthesis of graphene on h-BN *via* diffusion of catalytically produced carbon radicals through polycrystalline metal grain-boundaries and their crystallization at the interface of metal film and h-BN. The directly-grown, van der Waals bound graphene/h-BN heterostructures with atomically-precise, sharp, and contamination-free interfaces are obtained by carefully removing the top graphene and thin metal layers. The structural integrity, atomic registry, crystal quality and orientations of graphene/h-BN stacks are typified by scanning resonant Raman spectroscopy, X-ray photoelectron spectroscopy, transmission electron microscopy, and atomic force microscopy. Further, the low-temperature quantum transport measurements reveal the charge carrier mobility of 100 $cm^2v^{-1}s^{-1}$. Futuristically, this all-CVD direct-growth strategy will be a transformative approach for scalable production of van der Waals heterostructures to realize complex 2D circuitry.

524145 Overcoming Micro-Silicon Particle Fracture within Graphene Cages for Stable Battery Anodes

Abstract:

Silicon (Si) has over ten times the theoretical charge capacity of commercial graphite anodes, making it a promising material for enabling next-generation lithium-ion batteries. Nanostructuring has been shown to be fruitful in addressing the problems caused by large Si volume changes (~300%) during battery operation, including particle fracture, loss of electrical contact, and an unstable moving interface. However, nanoscale materials introduce new issues of higher cost and increased surface area for parasitic side reactions, resulting in poor Coulombic efficiencies and challenges for scaling up. Microscale Si powders are low-cost alternatives but, unlike Si nanoparticles, will inevitably fracture and lose electrical contact during charge/discharge. Thus, stable cycling of Si microparticles (SiMP) in batteries have been widely considered to be impossible.

In this work, we develop a method to encapsulate SiMP (diameter ~1-3 Î¼m) within conformally synthesized cages of multilayered graphene [1]. Despite volume expansion and particle fracture, the Si powders remain electrically connected on both the particle and electrode level by the mechanically robust and electrically conductive graphene cage. Furthermore, the chemically inert graphene cage forms a stable interface with the electrolyte. This minimizes irreversible consumption of lithium ions and increases the first-cycle Coulombic efficiency to as high as 93%, approaching the value of commercial graphite (~95%). With these advantages provided by the graphene cage, we demonstrate that even in a full-cell electrochemical test with a finite source of lithium ions, stable cycling (100 cycles; 90% capacity retention) is achieved for the previously nonfunctional microscale Si, bringing Si-based battery anodes one step closer to reality.

[1] Y. Li*, K. Yan*, H.-W. Lee, Z. Lu, N. Liu, Y. Cui. â€œGrowth of conformal graphene cages on micrometre-sized silicon particles as stable battery anodes,â€ *Nature Energy* 1 (2016): 15029.

*Denotes equal contribution

531976 Patterning of Defect-Engineered Graphene Sheets Driven By Pore-Pore Interactions

Abstract:

Defect engineering of graphene is a promising approach toward fabrication of carbon-based 2D materials with unique properties and function. Structural defects such as vacancy clusters (or nanopores) in graphene sheets, generated by electron or ion irradiation and placed accurately at preselected positions with almost atomic precision, play a key role in modifying the structure of graphene sheets and tuning their electronic, mechanical, and thermal properties. At high temperatures, under the action of thermodynamic driving forces, these defects exhibit interesting dynamics such as migration and coalescence, edge reconstruction of graphene sheets and graphene nanoribbons (GNRs), and GNR edge patterning mediated by nanopore-edge attractive interactions. These defect dynamical phenomena are of particular interest for developing graphene sheet patterning strategies. A fundamental understanding of defect interactions in graphene sheets is required to enable defect-mediated patterning strategies that can be used to tune 2D material properties and optimize their functionality.

Toward this end, we report here the results of a systematic atomic-scale computational analysis of nanopore dynamics in graphene sheets. The analysis is based on molecular-dynamics (MD) simulations according to a reliable interatomic interaction potential. The MD simulations are designed based on the results of a systematic protocol of molecular-statics (MS) computations to study the energetics of pore-pore interactions in graphene. The MS analysis reveals that when two pore edges are in close proximity (a few bond lengths) from each other, their interactions are attractive when the line segment connecting the two pore centers is aligned with the armchair direction. However, when this pore center-to-center line is along the zigzag direction, the interactions are both repulsive and attractive. Nevertheless, the repulsive interactions are very weak, while the attractive interactions become particularly strong as the pore centers approach closer than a certain distance. In both of these cases, pore-pore interactions can lead to pore coalescence and formation of a much larger pore in the graphene sheet. We have examined systematically the effects of pore sizes and other geometrical parameters on the pore-pore interaction potentials. With this understanding of pore-pore interactions, we have conducted molecular-dynamics (MD) simulations at high temperature to understand nanopore dynamics in the vicinity of other pores in graphene sheets. Our simulation results show that pore-pore interactions provide the thermodynamic driving force for a nanopore to migrate toward a larger pore, through a sequence of carbon ring reconstructions and edge atomic diffusion processes, and facilitate pore coalescence. A systematic parametric study varying pore geometrical parameters and temperature has identified four pore coalescence mechanisms. Based on analysis of MD trajectories and climbing-image nudged elastic band (NEB) calculations, we have constructed the minimum-energy pathways of these pore coalescence mechanisms and obtained a fundamental kinetic understanding of the processes underlying pore migration and coalescence in graphene.

534103 Synthesis and Applications of Magnetic Au-Ag-Î^{3}-Fe_2O_3 Nanocomposites on Reduced Graphene Oxide

Abstract:

Over the years, nanoparticles have been utilized widely in the fields of energy, biomedicine and environmental pollution prevention. Compared with bulk materials, noble metal nanoparticles are applied more broadly to heterogeneous catalytic processes because of their distinctive physicochemical properties, such as large specific surface area and high Fermi potential. Especially, polymetallic nanoparticles show enhanced catalytic properties caused by synergistic effect. However, the aggregation of nanoparticles may result in the decrease of catalytic activity. To maximizing the activity of noble metal nanoparticles, a suitable support is necessary for their dispersion. Graphene is a promising candidate which has attracted great attention due to its excellent chemical and physical properties, such as large specific surface area, outstanding mechanical strength, and unique thermal and electronic properties. As a derivative of graphene, graphene oxide becomes a common choice for support. It has unique properties just as graphene and is convenient to prepare. It should be noted that the nanocatalysts are difficult to use in continuous flow systems due to their tiny size and high stability in reaction systems. To solve this problem, nanocatalysts are combined with magnetic nanoparticles such as Fe_3O_4 and Î^{3}-Fe_2O_3. These hybrids are easy to retrieve from the reaction system under a magnetic field due to their magnetic properties. Magnetic separate technology has been widely used in separate fields and is quite desirable in industries because it can overcome the disadvantages of batch operation and other issues in filtration or centrifugation. In this paper, a facile route has been developed to prepare magnetic trimetallic Au-Ag-Î^{3}-Fe_2O_3/rGO nanocomposites. The nanocomposites were characterized by transmission electron microscopy (TEM), X-ray diffraction (XRD) and X-ray photoelectron spectroscopy (XPS). The prepared nanocomposites show fine catalytic activity towards the reduction reaction of 4-nitrophenol. The nanocomposites also have superparamagnetism at room temperature, which can be easily separated from the reaction systems by applying an external magnetic field. The multiplexing experiments were also preformed giving remarkable reusability and stability.

534632 Characterization of Lipid Dynamics and Structure on Epitaxial Graphene

Abstract:

Graphene is a two-dimensional material composed of a single carbon layer that offers several appealing properties including high conductivity, large surface area, and flexibility. Its unique properties make graphene an ideal substrate for several applications, including energy storage, optical electronics, and medical devices. Functionalizing graphene with lipids both increases its biocompatibility and provides a platform for diverse bioelectrochemical systems. However, characterization of lipids on graphene is challenging since traditional fluorescent methods for characterization of supported lipid structures are ineffective on graphene due to its highly quenching nature. Furthermore, there are multiple conflicting models published for the structure of lipids on graphene, and no standard approach for analyzing lipid properties on different types of graphene. We demonstrate that utilizing a modified Raman spectroscopy setup allows for characterization of lipids on graphene while providing additional benefits over conventional setups. We use this technique in conjunction with simulations to determine the structure, fluidity, and homogeneity of lipids on graphene.

537203 Detailed Characterization and Fabrication of 3D Printed Graphene/Polymer Structures for Heterojunction-Devices with MoS2 and Other 2D Nanomaterials

Abstract:

It is known that the 2D nanomaterials (2DN) can be interfaced with another to realize heterostructures with controlled and/or expanded properties, which can be applied to achieve functional devices. Currently, such structures has been produced at nano- to micro-scale dimensions. Here we show that 3D printing of graphene/polymer composite can allow device fabrication with 2DN-heterostructures produced at larger and accessible scale, while leveraging their unique properties. For example, fusing graphene, a zero bandgap 2DN and MoS_2, a semiconductor 2DN, together can produce ultrafast optoelectronic switches because of the mismatched Fermi level. Graphene/polymer-MoS2-graphene/polymer junctions were produced via 3D printing and experiments were conducted to measure the carrier transport characteristics. The results from carrier-transport studied at cryogenic temperatures (25 K) to room temperature were analyzed to determine transport barriers, carrier concentrations, Coulomb blockade, inter-particle capacitance, conduction mechanism and opto-electronic response. Furthermore, the effects of the mechanical strain in the 3D-printed heterostructure was investigated. We envision that these 3D printed structures with 2DNs will lead to an evolution of next-generation optoelectronic, electronics and electromechanical devices, which would be printed on-demand.

537659 Laser Induced Graphene Conductive Films

Abstract:

In this work, we developed a process that starts with raw coal and ends with high quality, transparent, patterned electronic circuits. Coal was treated to remove toxic components while isolating carbon materials that were refined into graphene and graphene oxide powders and films. TEM, XRD and Raman spectroscopy were performed on powders treated over a large thermal range (200 - 2500C). Transparent films were then formed via an ultrasonic spray system and characterized for optical transmission across a spectral range from the UV to the NIR. Thin films of graphene oxide were selectively patterned using a high-power laser to transform the material into highly conductive channels while a high resolution 2D Raman spectral map was taken *in situ*. Correlations between electrical and optical properties were obtained from conductivity measurements. The carbon-derived materials were slightly less transmissive than ITO in the UV but outperformed it in the NIR while having similar conductivities.

520319 Controlled Synthesis of Graphite Oxide: Formation Process, Oxidation Kinetics, and Optimized Conditions

Abstract:

Graphite oxide (GO) is one of the most extensively studied materials and has been tested for numerous applications due to its unique properties. Nevertheless, a quantitative description of complicated transfer and reaction rates of the oxidation of graphite is still not available, which essentially hinders large-scale production of high-quality GO and other graphene related materials. Herein, GO is prepared from flake graphite (FG) by a modified Hummers method. Size and color evolutions of time-dependent complementary domains, i.e. the starting GO domain and the remaining FG domain, are monitored during the entire oxidation process. The oxidation of FG in acidic oxidizing medium can be interpreted as a contracting area process controlled by phase boundary. Reaction rate constant (k) and kinetic parameters (E_a, lnA) are obtained after a systematic investigation of the influence of reaction temperature, concentration of sulfuric acid, and oxidizing agent on the oxidation, with the aid of R2 mathematical model. Using the kinetic results, the oxidation process and the conversion rate from FG to GO thus becomes controllable. Both intermediate products and final GO products are studied using XRD, TGA, and XPS to reveal the oxidation mechanism. The degree of oxidation of FG and the content of major functional groups on the surface of products can thus be optimized. Our research results are valuable to understanding fundamental mechanism of graphite oxidation and to achieving industrial production of high-quality GO with controllable degree of oxidation and tunable proportion of oxygen-containing functional groups for a variety of applications.

537122 WS_2-Induced Enhanced Optical Absorption and Efficiency in Graphene/Silicon Heterojunction Photovoltaic Cells

Abstract:

The Van Hove singularity (VHS) induced enhancement of visible-frequency-absorption in atomically-thin two-dimensional (2D) crystals provides an opportunity for improved light management in photovoltaics; however, it requires the 2D nanomaterial to be in close vicinity to a photojunction. In this report, we design a Schottky junction-based photovoltaic system with single-layer graphene atop n-type silicon (n-Si), which when interfaced directly with few-layers of tungsten disulfide (WS_2) with an architecture of WS_2-Graphene/n-Si *via* a bottom-up CVD synthesis strategy, enhanced its power conversion efficiency. Here, the WS_2 induced photon absorption, only one atom above the photo-junction enhanced short-circuit current density, and the reconfiguration of the energy band structure led effective built-in electric field induced charge carrier transport (enhanced open-circuit voltage, V_{OC}). Similar to graphene/n-Si Schottky junction, the WS_2-Graphene/n-Si double junction exhibited non-linear current density-voltage (J-V) characteristics with a 4-fold increase in J_{SC}(2.28 mA/cm^2 in comparison to 0.52 mA/cm^2 for graphene/n-Si) and 40% increase in the V_{OC} (184 mV compared to 130 mV for graphene/n-Si) with a 6-fold increase in the solar cell power conversion efficiency. Futuristically, we envision an evolution in 2D heterojunctions with sharp-transitions in properties within a few nanometers enabling control on optical-absorption, carrier-distribution, and band-structure for applications including tandem photovoltaic cells and 2D optoelectronic circuit-switches.

531676 Manufacturing Functional Membranes from Nanostructured Polymers

Abstract:

<u>**Manufacturing Functional Membranes from Nanostructured Polymers**</u>

William A. Phillip

Department of Chemical & Biomolecular Engineering

University of Notre Dame

Abstract:

Responsible management of the worldâ€™s water resources is essential to supporting human life on earth and the growing global population will only increase the demand for clean water resources. The success of seawater desalination by reverse osmosis highlights the large role membrane separations will play in helping society meet the demand for fresh water. In this talk, a next generation membrane platform fabricated from self-assembled block polymer precursors that can be modified post-assembly, in a facile and scalable manner, will be discussed. Most state-of-the-art membranes utilize a size-selective, steric exclusion mechanism to filter dissolved solutes from solution. In this regard, self-assembled block polymers are a promising platform for producing high-performance membranes because large areas of membrane that contain a high density of well-defined nanoscale pores can be readily produced using the self-assembly and non-solvent induced phase separation (SNIPS) technique. Furthermore, the performance profile of the block polymer membranes can be controlled through clever design of the macromolecular precursors. The performance of these membranes is pushing the limits of size-selective separation mechanisms, which is driving interest in chemically-selective membranes that allow for efficient separations based on chemical factors. As such, block polymer membranes with nanoscale pores lined by moieties that can be easily reacted to other functionalities further demonstrate the promise of this platform. In one example, the pore-lining groups are converted to coordinating groups that act as high capacity binding sites for heavy metal ion adsorption. In another example, chemically-patterned charge mosaic membranes, are prepared by modifying the pore wall chemistry of the copolymer membrane precursors using ink-jet printing devices. The well-defined counter-charged domains that cover the surface of the resulting charge mosaic membranes allow dissolved salts to permeate more rapidly than waterâ€"even though water is three times smaller in size. Through these examples, we will demonstrate that membranes that are based on self-assembled block polymer materials provide a scalable platform that can be tailored to myriad separations for the purification and conservation of fresh water resources.

Â

534818 Electron Tomography Reveals Details of the Internal Microstructure of Desalination Membranes

Abstract:

As water availability becomes a growing challenge in various regions throughout the world, desalination and wastewater reclamation through technologies such as reverse osmosis (RO) are becoming more important. Nevertheless, many open questions remain regarding the internal structure of thin-film composite RO membranes. In this work, fully aromatic polyamide films that serve as the active layer of state of the art water filtration modules were investigated using high angle annular dark field scanning transmission electron microscopy (HAADF-STEM) tomography. Reconstructions of the 3D morphology reveal intricate aspects of the complex microstructure not visible from 2D projections. We find that the polyamide top surface areas, believed to be a major contributor to the overall flux of RO membranes, normalized to projected areas are 3.48 ± 0.15 and 3.07 ± 0.04 for the seawater and brackish water desalination membranes, respectively, which are more than double the values measured by techniques such as atomic force microscopy (AFM). This is because of the limited accessibility of the AFM tip to the overall highly heterogonous and three-dimensional polyamide top surface. Furthermore, the internal voids of the active layer of compressed commercial membranes account for less than 0.2% of the total polymer volume, contrary to previously reported values that are two orders of magnitude higher. Thus, 3D reconstructions and quantitative analyses will be crucial to characterize the complex morphology of polymeric membranes used in next-generation water purification membranes.

536731 Synthesis and Self-Assembly of a New High-χ Block Copolymer: Ptbs-b-Phema

Abstract:

Block copolymers (BCPs) have the ability to microphase separate into a variety of periodic morphologies (e.g. spheres, cylinders, lamellae) with a feature-to-feature spacing (pitch) ranging from 5 to 100 nm. In particular, cylindrical and lamellae morphologies are of interest to the microelectronics industry since they can be used as creating masks for nanofabrication of contact holes and line-space patterns, respectively. As the industry tackles the challenges of continuing Moore's Law, BCPs offer an alternative route to more expensive pathways such as Extreme Ultra-Violet Light lithography (EUVL). The pitch of a BCP is dependent on both the interaction parameter (χ) and the degree of polymerization (*N*), with a much stronger dependence on the degree of polymerization. Therefore, in order to reach small feature sizes and pitches in the microphase block copolymer assembly, a low *N* and a high χ are required. Ideally a BCP will be coated on a substrate and raised above its glass transition temperature to microphase separate until a defect free state is reached. By using chemically preferential patterned underlayers, these features can undergo directed self-assembly (DSA) where by their features orient in particular directions rather than random. By removing of one of the BCP's blocks, an etch mask of line-space or contact holes can be produced over a substrate.

Here we report on the synthesis and characterization of a new high-χ BCP, poly(4-tertbutylstyene)-b-poly(2-hydroxyethylmethacrylate) (PtBS-b-PHEMA). The χ for PtBS-b-PHEMA is expected to be large due to the hydrophobic PtBS and hydrophilic PHEMA having a high penalty for interacting with one another. High χ materials not only possess the ability to reach small pitches but they also have a high penalty of defect formation, as any defects lead to increased interaction area between the blocks and an enthalpic penalty for the system. The BCP was synthesized by anionic polymerization of a TBDMS protected hydroxyethyl methacrylate monomer in conjunction with a t-butyl styrene monomer. Small angle X-ray scattering (SAXS) of the bulk BCP shows that it is capable of phase separating into various morphologies (lamellae, cylinders, and gyroids) exhibiting pitches less than 7nm. Investigation into a neutral underlayer for thin films of PtBS-b-PHEMA was also conducted using photo-definable random copolymers with the ability to modulate their surface energy. Thin film self-assembled morphologies were produced via thermal annealing and viewed using SEM and AFM. An overview of this new block copolymer system and the associated materials required to produce DSA patterns of the material will be reviewed.

529893 Assembly and Photoswitching Dynamics in Nanostructured Polymer Thin Films Revealed By Single-Molecule Super-Resolution Microscopy

Abstract:

Single-molecule super-resolution microscopy is attracting increasing attention in materials science by offering an exciting method for direct *in situ* real-space observation at nanometer resolutions. In this work, we develop a photoswitchable fluorophore optimized for organic environments, and we explore its incorporation into a super-resolution microscopy experiment. The fluorophore is polymerizable which greatly streamlines labeling strategies into a single step, where a small quantity is mixed with standard monomers and copolymerized to yield the functionalized material with no further purification. Individual fluorophores can be switched several times between bright and dark states, enabling repeated time-lapse imaging. We explore the photoswitching kinetics by comparing to Monte Carlo simulations, suggesting that the reaction dynamics are heterogeneous. *In situ* nanoscale imaging is demonstrated in solvent-swollen polymer blends, where coarsening behavior is readily observed. We hope our imaging platform promotes widespread adoption of super-resolution microscopy in the polymer community.

537175 Nanostructured Polymer Gels and Brushes Via 2 Color Interference Lithography

Abstract:

Optical interference lithography has remained an attractive technique to cheaply and rapidly pattern three dimensional features in polymer photoresists despite both resolution and feature size being limited by diffraction. In the past few years, Stimulated Emission Depletion Microscopy (STED) inspired lithography schemes have shown the ability to direct-write features well below the diffraction limit using visible light. However, the high intensity required for effective photoinhibition limits their use to point by point writing.

Recently, we have shown that combining the reversibly saturable photoisomerization of spirothiopyran with thiol-Michael addition chemistry can be used to formulate writing systems with desired low threshold characteristics. Adapting the spirothiopyran writing chemistry for polyethylene glycol monomers allows us to synthesize photoresponsive gels where crosslinking can be initiated and inhibited using two different wavelengths of light with spatial control obtained by controlling the relative intensity of the activating and deactivating wavelengths. Photokinetic modeling of such systems demonstrate the potential to fabricate nanostructured gels with feature sizes less than 50 nm over ranges spanning hundreds of microns by tuning the kinetics of spirothiopyran photoswitching using a 2W 532 nm light source. Nanopatterning using spirothiopyran functionalized systems is experimentally demonstrated by spatially localizing the Michael addition product over 1D patterns spanning large areas (>500 μm^2) using a 2-color interference lithography setup. The feature size of the written patterns is shown to be tunable by controlling the relative intensity of the initiation and inhibition wavelengths. Secondary patterning on such nanostructured gels using unreacted photochromic brushes in conjunction with bifunctional maleimide linker molecules can be used to tailor the patterned surfaces to desired chemical functionalities. These experiments demonstrate the versatility of spirothiopyran based photochromic polymer systems for fabricating large area chemically tunable nanostructured gels

535473 Effect of Curing Bath Conditions on the Morphology of Porous Hollow Poly(High Internal Phase Emulsion) Fibers

Abstract:

Porous hollow fibers have a variety of important technological applications including water treatment, the semiconductor industry, and also biomedical uses. Conventional methods for fabricating porous hollow fibers generally rely on electrospinning which requires the use of toxic organic solvents. Fabricating porous hollow fiber via templating from high internal phase emulsions (HIPEs) provides an environmentally friendly process alternative. In this work, we demonstrate how manipulation of the composition of the bath in which the poly(HIPE) fibers are cured can be used to tune the macroscopic and microscopic structure of the porous hollow fibers.

To create the porous hollow fibers, HIPE is extruded into an aqueous curing batch containing initiator and electrolyte (NaCl) in various concentrations. Through the combination of osmotic pressure effects (manipulated through setting the NaCl concentration difference between the curing bath and the aqueous phase within the HIPE) and the initiator concentration (which triggers the polymerization reaction at the HIPE-bath interface), different hollow fiber morphologies can be created. Scanning electron microscopy(SEM) image shows that parameters such as initiator concentration, electrolyte (NaCl) concentration, and curing temperature can be used to adjust the size of the hollow center and wall thickness, as well as the microstructure of the porous wall. This new approach enables the possibility of designing the morphology and structure of porous hollow fiber via manipulation of curing bath and processing conditions.

535766 Geometry and Composition of Soft Polymer Films Embedded with Nanoparticles Enhance Rates for Optothermal Heat Dissipation

Abstract:

Embedding soft matter with nanoparticles (NPs) can provide electromagnetic tunability at sub-micron scales for growing number of uses in biomedicine including image contrast agents, biosensors, tissue culture substrates, photothermal therapy, drug delivery, and gene transfection. But use of NP-embedded soft material in temperature-sensitive applications has been constrained by difficulty in validating prediction of rates for energy dissipation across thermally insulating to conducting behavior. For example, it has been shown that increasing NP concentration in photothermal therapy paradoxically resulted in tumor regrowth despite increased temperatures, because of reduced depth of laser penetration.

To date, rigorous evaluation of NP-induced photothermal effects in biological systems has been largely limited to characterizing steady state heat flux or temperature distributions. However, the (dynamic) rate of temperature increase, i.e., heating rate, in such systems is largely unexplored. The rate of heat dissipation is a critical factor influencing a variety of biomedical outcomes. For example, a heat rate not less than one degree celsius per minute is the European Society for Hyperthermic Oncology guideline for superficial hyperthermia clinical trials. In NP-containing materials, heating rate was reported to vary nearly 3-fold depending on the configuration of the NP. And heating transients have been shown to influence rate of drug release from polydimethylsiloxane (PDMS) films and phase change materials.

The present work examined changes in the transient rate for heat dissipation as inter-NP separation was reduced from greater than 5 microns to sub-micron intervals while varying the 3D geometry of a polymer and its NP density 2- to 15-fold.To our knowledge this is the first quantitative comparison of simulated and experimentally measured plasmon-thermal heating transients that spans thermally conductive to thermally insulative regimes. In order to achieve this result, embedment of monodisperse NP was improved using a refined fabrication process to stably decrease inter-NP spacings in biocompatible polymer to nano-scales. Lumped-parameter and finite element analyses were refined to apportion effects of structure and composition of NP-embedded soft polymer to rates for conductive, convective, and radiative heat dissipation. These advances allowed rational selection of PDMS size and NP composition to optimize measured rates of internal (conductive) and external (convective and radiative) heat dissipation.

Stably reducing distance between monodisperse NP to nano-scale intervals increased overall heat dissipation rate by up to 29%. Refined fabrication of NP-embedded polymer enabled tunability of dynamic thermal response, the ratio of internal to external dissipation rate, by a factor of 3.1 to achieve a value of 0.091, the largest reported to date. Heat dissipation rates simulated *a priori* were consistent with 130 µm resolution thermal images across 2- to 15-fold changes in NP-PDMS geometry and composition. Nusselt number was observed to increase with the fourth root of Rayleigh number across thermally insulative and conductive regimes, further validating the approach. These developments support model-informed design of soft media embedded with nano-scale spaced NP to optimize heat dissipation rates for evolving temperature-sensitive diagnostic and therapeutic modalities as well as emerging uses in flexible bioelectronics, cell and tissue culture, and solar-thermal heating.

525701 Invited Speaker: Lipophilically-Functionalized Porous Silica Nanoparticles for Acoustic Imaging and Site-Specific Therapy

Abstract:

Ultrasound is widely applied in medical diagnosis and therapy due to its safety, high penetration depth, and low cost. In order to improve the contrast of sonographs and efficiency of ultrasound therapy, echogenic gas bodies or droplets (with diameters from 200 nm to 10 μm) are often used. However, because their inherent Laplace pressure limits both size and stability these fluid-filled colloids, they are not stable in circulation and have difficulty penetrating effectively into target tissues. Here we will present our work designing silica nanoparticles of ~100 nm with specifically-tailored surfaces that can nucleate the formation of ultrasound-responsive microbubbles under reduced acoustic pressures. This talk will discuss three aspects of this work. First, a comprehensive study was undertaken to determine the effect of structure on bubble nucleation and cavitation. A series of mesoporous silica nanoparticles were synthesized with sizes around 100 nm, each containing different morphologies. From these studies, the effects of nanoparticle porosity, surface roughness, hydrophobicity, and hydrophilic surface modification on acoustic cavitation inception by porous nanoparticles were determined. Second, particles with a hexagonal, small pore morphology were utilized for imaging in complex media while resisting protein fouling. Because the combination of surface roughness and hydrophobicity was necessary for nucleating bubble cavitation, proteins such as albumin adhering to the surface of the nanoparticle would be expected to quench signal entirely. The nanoparticles were found to provide effective image contrast even in whole blood and long-term storage. Finally, the therapeutic efficacy of the particles was determined by utilizing a phospholipid monolayer that was functionalized with PEG-lipid. The nanoparticles were uptaken into cancer cells by receptor-mediated endocytosis, and the cells with uptaken particles were killed selectively under HIFU insonation, as compared to cells without nanoparticles. Thus, these studies show how small nanoparticles can sensitize both bubble formation and cell death, thus overcoming the inherent Laplace limit found with fluid-filled particles.

537166 Nanoparticles for Combination Chemotherapy

Abstract:

Nanoparticles for Combination Chemotherapy

Shani L. Levit, Christina Tang

School of Engineering, Department of Chemical and Life Sciences Engineering

The effectiveness of chemotherapy (e.g. paclitaxel) can decrease with continual use due to development of multi-drug resistance (MDR). MDR results in an overexpression of ATP-binding cassette (ABC) transporters, which remove the anti-cancer drugs from the tumor cells. ABC transporter inhibitors, e.g. lapatinib, have be used in combination with chemotherapeutics (e.g. paclitaxel) to overcome MDR. Current practices include sequential delivery of lapatinib followed by paclitaxel. However, these drugs are poorly soluble in the blood and cause severe side effects, limiting their efficacy. pH-responsive nanoparticles could facilitate controlled drug release to increase drug efficacy while limiting side effects. One approach to achieving pH responsive nanoparticles has been encapsulation of anti-oxidant metal complexes using a block copolymer stabilizer via Flash NanoPrecipitation (FNP). In FNP, an antioxidant, tannic acid (TA), dissolved with an amphiphilic PEG-based block copolymer is rapidly mixed with Fe^{3+} (aq.). Upon mixing at pH 7.4, TA and Fe^{3+} form insoluble complexes which are encapsulated and stabilized by the amphiphilic block copolymer. The resulting nanoparticle is pH-responsive; the nanoparticles are stable under basic conditions (pH >7) when the complex is insoluble and disassemble in acidic environments (pH <2) when the complex is soluble. The overall goal is to achieve controlled release of combinations chemotherapies. In this work, we encapsulate weakly hydrophobic drugs (i.e. paclitaxel, lapatinib, and sulfasalazine) using two methods 1) direct encapsulation by dissolving the drugs with TA and the block copolymer and 2) encapsulation of hydrophobic prodrugs. For example, particles encapsulating paclitaxel 128 nm ± 3 nm in diameter have been achieved via direct encapsulation. The encapsulation efficiency of paclitaxel was ~ 80%. The size and encapsulation efficiency will be compared for single and co-loaded nanoparticles. The effect of pH on drug release will be discussed.

537766 Co-Loading of Hydrophilic and Lipophilic Therapeutics through Equilibration within Temperature Sensitive Liposomes

Abstract:

Introduction: Since their discovery nearly five decades ago, liposomes have become the most well studied nanodelivery system. Researchers have exploited their amphipathic nature to sequester both lipophilic and hydrophilic compounds in order to make potent biodegradable and biocompatible delivery vehicles. However, despite their well-documented use, strategies for encapsulating therapeutics are limited and are often extremely specialized, hindering their widespread implementation. Several techniques, including passive and active loading strategies, have been proposed to formulate potent drug encapsulated liposomes. Passive loading has been recognized as a universal encapsulation technique, but often suffers from poor loading efficiencies. Conversely, active loading is extremely efficient, but is limited to weakly basic therapeutics, making it dependent on the chemical nature of the drug. Our work aims to build upon our SPIN (synthesis and purification of injectable nanocarriers) technology for liposome synthesis and develop an efficient and universal loading strategy. A novel equilibration technique is used as a one-step approach to co-load both lipophilic and hydrophilic molecules, simultaneously. We demonstrate the efficacy of co-loaded liposomes by encapsulating two front-line chemotherapeutics, Doxorubicin (DXR) and Paclitaxel (PTX), and measuring their cytotoxic effect on MDA-MB-231 metastatic breast cancer cells. To understand the atomic level interactions that occur between DXR, PTX, and the liposomal membrane during equilibration, we perform molecular dynamic simulations using NAMD (Nanoscale Molecular Dynamics). We believe that this scalable and robust technique for synthesizing potent co-loaded liposomes offers unparalleled advantages compared to contemporary encapsulation techniques, and will ultimately lead to the development of novel liposomal formulations for clinical intervention.

Materials and Methods: Liposomes composed of 1,2-Distearoyl-sn-glycero-3-phosphorylcholine (DSPC) and cholesterol at a 2:1 ratio are synthesized using our SPIN technique. The synthesized liposomes are equilibrated at the transition temperature (T_m) of DSPC (55 °C) for up to 4 hours with DXR and PTX. The co-loaded liposomes are then filtered to remove non-encapsulated therapeutics and the drug to lipid ratio of the two molecules in the liposomes is calculated. Samples were incubated with MDA-MB-231 metastatic cancer cells and viability over time was analyzed using a WST-1 proliferation assay. All-atomic NAMD simulations were performed using the CHARMM force-field for both lipids and drugs. Simulations were conducted for 100 ns at temperatures of 37 ^{O}C, 55 ^{O}C, and 65 ^{O}C.

Results and Discussion: Liposomes were produced using our SPIN technique, which is an efficient and scalable synthesis approach. Using equilibration, we were able to achieve drug to lipid ratios of 0.04. Cells incubated with liposomal therapeutics showed a drastic decrease in viability over time as compared to controls. NAMD simulations to investigate the interactions between the drugs and the membrane show that electrostatic forces play a predominant role in the encapsulation of therapeutics.

537308 Novel Nano Biotechnology Approaches for Treating Intracellular Bacterial Infections

Abstract:

Bacterial infections were once easily treatable illnesses but are increasingly becoming a serious world health epidemic. Rising antibiotic resistance is making current treatments obsolete and even effective antibiotics are inadequate when bacteria invade the hostâ€™s cells and create a persistent or chronic infection.

To combat an intracellular infection, we developed two novel antimicrobial treatments: light activated nanoparticles and gene-specific antisense molecules. Light activated cadmium telluride (CdTe) quantum dots are precisely tuned nanoparticles that treat infectious agents by increasing the reactive oxygen species (ROS) content exposed to the bacteria to its lethal threshold while remaining below the host cellsâ€™ toxic ROS limit. Due to the CdTe nanoparticles 2-4 nm diameter they are small enough to penetrate the host cell membrane, bacteria vacuole, and bacteria membrane to easily treat intracellular infections. Secondly, here we show a gene-specific antisense therapy using peptide nucleic acids (PNA) can offer a precise method of targeting and treating intracellular infections. PNA therapeutics can be designed to target numerous essential genes unique to a specific bacterium of interest without effecting non-infectious bacteria.

Here we establish an infection model using *Salmonella enterica* serovar Typhimurium in two mammalian cells lines, the epithelial HeLa cell line and osteoblast precursor MC3T3-E1 cell line, to test our two novel antimicrobial treatments and their effect on the host cells. *Salmonella* is a common microbe seen in infections that readily infects numerous different cells lines using its type three secretion system (T3SS) and remains intracellular in a protective salmonella containing vacuole. Exogenously added quantum dots show tunable clearance and limited host cell toxicity by modulating the concentration and light intensity, correlating to the quantum dotsâ€™ ROS production. PNA therapeutic molecules when added using bacterial secretion systems, also show tunable intracellular clearance based on concentration and gene specificity with limited host cell toxicity.

These results demonstrate two new antimicrobial treatments to combat intracellular infections and establish an infection model to further explore these therapeutics on different combinations of intracellular infections and host cells representing various pertinent clinical infection challenges.

536648 A Virus-Free Fe_3O_4 Nanoparticle-Based H7N9 Influenza Vaccine

Abstract:

One hundred years has passed since the A/H1N1/1918 influenza caused millions of deaths worldwide. The avian H7N9 influenza virus that emerged in China in 2013 showed a mortality rate approaching 40% with no vaccine available at the time. These influenza outbreaks emphasize the need for rapid manufacture of influenza vaccines for a fast response to epidemic emergencies.

In this contribution, we integrate the use of recombinant technology and nanotechnology to synthetize an influenza nanovaccine candidate. We engineered an *Escherichia coli* BL21 C41 strain to produce the globular region of the hemagglutinin (HA) from the H7N9 virus in bacteria. The recombinant protein was recovered and purified from the insoluble fraction of the cellular lysate using histidine tag affinity chromatography. The recovered HA protein was attached to Fe_3O_4 paramagnetic nanoparticles functionalized with NH_2 groups through EDC coupling in a microfluidic device. The resulting nanoparticles exhibited anti-HA activity in ELISA experiments when tested against H7N9 polyclonal antibodies. In addition, this nanovaccine candidate triggered a systemic immune response (measured as an increased IgG plasma concentration) and elicited a specific immune response in a chicken model immunized with doses of ~5 μg/kg of body weight (as determined by ELISA).

This class of virus-free vaccine candidates may represent a low-risk and cost-effective alternative for animal and human anti-influenza vaccination.

538784 Engineering Lipid Nanoparticles to Mitigate Oxidative Stress in Stem Cell Transplant Therapy

Abstract:

The cell transplant and organ transplant therapies have been widely used for several applications in medicine. The integrity of the transplantation process is highly dependent on cell adhesion, migration and proliferation. However, the reactive oxygen species (ROS) generated in surroundings during this process inhibits the cell-cell and cell-matrix adhesion, affect cell spreading and lead to increased detachment of cells. Reactive oxygen species can also induce an inflammatory response thereby damaging the cellular structure and function and leading to cell death. Thus the survival rates of cells when transferred in vivo are very poor because of the oxidative stress that they experience. Pre-treatment of cells with antioxidant prior to the transplant process can help in increasing the survival rate of cells. Vitamin E can be used as a potential drug owing to its antioxidant properties. Targeted drug delivery using nanoparticles has several advantages over the traditional drug delivery method such as reduced dosage of drug, lesser side effects, decrease in cost of therapy and increased accumulation of drug within the target tissue. Liposomes are very biocompatible and highly tunable for specific applications. Thus the vitamin E loaded lipid nanoparticles can be efficiently used for targeting the site of oxidative stress. In this project, we are developing highly customizable lipid based nanoparticles encapsulating vitamin E for targeting the site of oxidative stress. We have engineered lipid nanoparticles (LNPs) of various sizes, surface charges, and coating density of hydrophilic polymer using non-cationic lipid and cholesterol components. LNPs were characterized by both photon correlation spectroscopy to determine particle size distribution, hydrodynamic diameter, and polydispersity, and by zeta potential measurements to determine nanoparticle surface charge characteristics and to quantify colloidal stability. The encapsulation studies were performed using the absorbance measurement of vitamin E at 295nm wavelength and we obtained upto 75 % of encapsulation efficiency. The preliminary uptake studies using stem cells and FITC tagged LNPs indicated that the nanocarriers were internalized into the cytosol. Then the stem cells were treated with peroxide (H202) at different concentrations to create an oxidative stress environment. The vitamin E loaded LNPs targeted to these cells showed a significant decrease in the reactive oxidative stress induced due to peroxide. We are currently examining the release kinetics of vitamin E from the LNPs. The future work of this research involves optimisation of the process along with the in-vivo studies of stem cell transplantation in mice.

536948 Sugar-Guided Organ and Cellular Targeting of PAMAM Dendrimers

Abstract:

Hydroxyl-terminated polyamidoamide (PAMAM) dendrimers have been proven to be effective, versatile nanocarriers for delivering therapeutics and imaging agents across the blood-brain barrier, where they are selectively taken up by injured microglia and macrophages. This passive targeting has been explored in great detail for the delivery of an anti-inflammatory compound, N-acetyl cysteine, to activated glia in many small and large animal models of central nervous system (CNS) disorders [1-3]. The generation 4 dendrimer scaffold (D4-OH), which is a 4nm diameter spheroid with 64 hydroxyl end groups, is known to be non-toxic and non-immunogenic [4]. The N-acetyl cysteine conjugate (D-NAC) has been shown to maintain that biocompatibility and is currently undergoing clinical translation for a pediatric neurological disorder.

Despite the advances that have been made with D4-OH, there remains immense untapped potential in applications from diagnostics to treatment due to the dendrimer's propensity for residence exclusively in injured glial cells and activated macrophages. Building on the unique biophysics of hydroxyl PAMAM dendrimers, we seek to further tailor the dendrimer's pharmacokinetics and transport properties by utilizing various sugars as targeting ligands to create a new catalog of dendrimer conjugates with the ability to precisely target different cells and organs throughout the body. Here we will discuss the synthesis of new sugar-dendrimer conjugates and their performance *in vitro* and *in vivo* compared to D4-OH. The conjugates of interest in this study are: dendrimer-galactose (D-Gal), dendrimer-glucose (D-Glu), and dendrimer-mannose (D-Man) (**Fig 1**).

We have observed that conjugation of any of the three sugars to the dendrimer surface results in stark deviations in dendrimer pharmacokinetics as analyzed through tissue homogenization and fluorescence spectroscopy. D4-OH has a diameter of about 4nm, which lends itself to rapid renal clearance; however, in the cases of both D-Gal and D-Man the attachment of just 12-14 sugar molecules to the dendrimer surface (<15 wt%, diameter~ 5 nm) allows the dendrimer to localize in the liver, which has not typically been observed to metabolize the nanoparticle. In a neonatal rabbit model of cerebral palsy, over 9% of the injected D-Man dose was present in the liver at 24 hours post-injection compared with D4-OH, of which only less than 1% of the injected dose had localized to the liver (**Fig 2**). Preliminary results indicate that compared to D4-OH in healthy mice, D-Gal levels are ~3-fold higher in the lungs and brain, ~2-fold higher in the kidneys, and over 10-fold higher in the liver at 24 hours post-injection.

These modifications to overall dendrimer biodistribution indicate that the conjugation of sugar to the dendrimer surface alters dendrimer-cell interactions and dendrimer transport, which would also effect the cellular localization of the dendrimer. We explored this phenomenon through *in vitro* competitive uptake and binding assays as well as confocal imaging and flow cytometry of *ex vivo* tissues. The conjugation of mannose to the dendrimer surface gave the dendrimer a partiality for cellular uptake through mannose receptor-mediated endocytosis, whereas D4-OH has been observed to utilize a variety of fluid-phase endocytic pathways over mannose receptor-mediated endocytosis [5]. This was confirmed in confocal imaging which found D-Man localized in mannose receptor-expressing cells, such as the meninges of the cortex and sinusoidal endothelial cells of the liver. Hepatocytes comprise almost 80% of the liver by mass, yet have remained an elusive target for nanoparticles; however, we have observed that D-Gal is highly effective at penetrating and remaining in hepatocytes, with strong dendrimer signal seen up to one week post injection in hepatocytes of healthy animals (**Fig 3**) most likely due to the high expression of galactose receptors, such as asialoglycoprotein receptor 1. D-Gal, D-Glu, and D-Man are also under investigation in a mouse model of glioblastoma to see if sugar conjugation allows the dendrimer to take advantage of the Warburg effect and localize in tumor cells as well as tumor-associated macrophages.

Additional changes to cellular localization in other organs and disease models are still being explored, but the results as such indicate that the conjugation of sugar to the dendrimer surface modifies its intrinsic transport properties, allowing the dendrimer to bind to different surface receptors, which has dramatic effects on the pharmacokinetics of the dendrimer. This new catalog of dendrimers gives new hope for treatment of hepatic diseases, CNS disorders, and diseases of the back of the eye, increasing the already expansive utility of this dendrimer nanoparticle.

This work was funded by grants from the National Institute of Biomedical Imaging and Bioengineering (5R01EB018306-02) and the National Institute of Child Health and Human Development (5R01HD076901-02).

[1] Kannan, et al. Dendrimer-based postnatal therapy for neuroinflammation and cerebral palsy in a rabbit model. *Sci. Trans. Med.* 2012. 4(130):130-46.

[2] Grimm, et al. Nanotechnology Approaches to Targeting Inflammation and Excitotoxicity in a Canine Model of Hypothermic Circulatory Arrest-Induced Brain Injury. *Ann Thorac. Surgery*. 2016. 102(3):743-750.

[3] Nance, et al. Dendrimer-mediated delivery of N-acetyl cysteine to microglia in a mouse model of Rett syndrome. *J Nueroinflammation*. 2017. 14(1):252.

[4] Dobrovolskaia, MA. Dendrimers Effects on the Immune System: Insights into Toxicity and Therapeutic Utility. *Curr Pharm Des*. 2017. 23(21):3134-3141.

[5] Alnasser, et al. Preferential and increased uptake of hydroxyl terminated PAMAM dendrimers by activated microglia in rabbit brain mixed glial cultures. *Molecules*. 2018. In press.

535029 Engineered Electrodes for Energy Storage and Battery Safety

Abstract:

The ViPER (Vilas Polâ€™s Energy Research) laboratory at Purdue University focuses research efforts on the development of high capacity electrode materials and their engineering with long cycle life and improved safety. Considering the advantages and limitations of known synthesis techniques, various processing technologies are developed to fabricate a variety of unique anode and cathode materials for lithium-ion, Na-ion, K-ion and Li-S batteries. ViPERâ€™s recent efforts on structural, morphological and electrochemical properties of various electro-chemistries will be demonstrated.

528867 Extended Surface Electrocatalyst Development Via Atomic Layer Deposition

Abstract:

Extended surface electrocatalysts based on nanowire templates have emerged as promising candidates for advanced catalysts for the oxygen reduction reaction (ORR) in fuel cell applications. When synthesized by spontaneous galvanic displacement (SGD) of nickel nanowires (NiNWs), these extended surface electrocatalysts have demonstrated exceptional performance (specific activity > 6000 Î¼A/cm^2) and superior durability to benchmark carbon-supported Pt catalysts. However, challenges in scale-up and reproducibility of this fabrication process have prompted an investigation into atomic layer deposition (ALD) as a facile, scalable synthesis alternative for the production of extended thin film electrocatalyst structures. An ALD analog of the SGD synthesis process was developed using an oxidative Pt ALD chemistry in a viscous flow reactor. In situ mass spectrometry was used to study the deposition reactions, while elemental analysis, transmission electron microscopy, elemental mapping and X-ray diffraction were used to characterize Pt growth and effects of post-treatment procedures, such as H_2 annealing. Electrochemical half-cell testing showed that the highest-performing ALD catalyst was 16 times more active than commercial Pt/C. After integration into membrane electrode assemblies, these materials were also found to exceed the DOE 2020 mass activity target of 440 mA/mg_{Pt}. Further work is also underway to leverage the tunable nature of ALD to develop a co-deposition process with Pt and Ni ALD on a cobalt nanowire (CoNW) substrate. Fabrication of these trimetallic extended surface catalysts using ALD â€œsupercyclesâ€ and subsequent characterization experiments will also be discussed.

534820 Thin Oxide Film Coatings for Improved Lithium Ion Battery Cathodes

Abstract:

The application of alumina ALD films to battery cathodes is known to enhance the cycling stability of lithium ion batteries. The true nature and deposition mechanism of these films is not well understood. In this work we will elucidate the nature of low cycle ALD films on NMC cathode materials and discuss the mechanistic causes behind the characteristics of the films. An ultra-thin alumina film is applied to cathode particles by atomic layer deposition (ALD) with a variety of cycle numbers. The films generated within the first 10 cycles of deposition are shown to improve the cycling stability of lithium ion battery cathodes. Low energy ion scattering (LEIS) and computational modeling were used to elucidate the coating mechanism that leads to the observed improved performance.

530256 The Application of Magnesiothermic Reduction of Silica to Produce Porous Silicon for Lithium Ion Batteries

Abstract:

Increasing demand for portable power applications are pushing conventional battery chemistries to their theoretic limit. Silicon has the potential as an anode material to increase lithium-ion cell capacity. The associated volume change of lithiation leads to a decline in capacity during cycling and low lithium diffusion rates within silicon limit high rate discharging. Many silicon morphologies and nanostructures have been studied including porous architectures. Porous morphologies can potentially address the poor cyclability and rate capabilities simultaneously. The template assisted synthesis â€˜magnesiothermic reductionâ€™ of silica to silicon offers a facile and scalable synthesis route to porous silicon structures. However, a review of the recent progress of this method indicates a distinct lack of mechanistic understanding of this phenomenon. To better understand the processes, we have systematically studied the process conditions using a model silica system. Further, we performed a complete characterisation of both the reactant and product structures in order to map how reaction conditions affect the properties of the electrode materials. Finally, we will present the battery performance of these porous silicon structures and compile a processing-structure-property-performance relationship.

539777 Fluidized-Bed CVD of Si@SiC@C-like Nanoparticle and Its Application As Anode Materials

Abstract:

Among the anode materials, silicon (Si) has the poor cycling performance induced by the large volume expansion (300%) which can break the solid electrolyte interface (SEI) during the lithium ion insertion/extraction, although it can provide the highest theoretical capacity as the most promising candidate for the next generation lithium ion batteries. For the purpose of make a stable SEI, a novel Si@SiC@C-like nanomaterial has been designed to gain 800mAh g^{-1} at the current dense of 1 A g^{-1} after 800 cycles. Besides, those nanoparticles can be easily obtained though chemical vapor deposition (CVD) in a fluidized bed at low cost. A reaction mechanism and the rate equations have been proposed based on the understanding of the effect of reaction temperature, partial pressure of reactant gas and time of synthesis on the yield of Si@SiC@C-like nanoparticles.

535397 Functional Nanomaterials for OIL and Gas Discovery and Recovery Applications

Abstract:

Today, the potential of functional nanoparticles and nanostructured materials to transform key upstream applications in the oil industry is evident. Functional nanoparticles can be engineered to traverse the reservoir while reporting their location, recording key reservoir rock and fluid parameters, and actively mobilizing oil. They enable target-delivery of chemical treatments and sensors deep in the reservoir and can be designed to migrate autonomously via harvesting the surrounding natural or induced solute or chemical gradients. Thus, they can substantially improve reservoir characterization and recovery of its target resources without excessive use of water and chemicals, and with minimal pumping energy.

The major challenges to maximizing the benefits of nanoparticle-based technologies in upstream oil applications can be summarized as: 1) lack of long-term stability and mobility of nanoparticles under typical oil reservoir conditions (usually temperatures exceeding 100 °C, salinities between 50,000 and 220,000 ppm, and pressures exceeding 3000 psi), 2) low-cost prerequisite due to the significantly large amounts needed to interrogate pore volumes typical to the reservoirs (millions of cubic meters), and 3) difficulty to precisely characterize the nanomaterials and test their performance under typical reservoir conditions.

In this article, we present three of our successful oil reservoir nanoagents; A-Dots, Nano-Surfactants, and Magnetic Nano-Mappers. The synthesis of these nanoagents and results of experiments to characterize their nature and behavior under typical reservoir conditions are presented.

A-Dots are carbon dots that are persistently stable at high salinity and elevated temperature. They are optically detectable at low concentrations and can migrate efficiently in the reservoir between injectors and producers. They thus allow establishing â€œcommunicationsâ€ between injectors and producers for improved management of the reservoir. Nano-surfactants are 10 to 60 nm nanostructured particles of the seawater-insoluble petroleum sulfonate salt surfactants, stabilized in seawater by by cocamidopropyl hydroxysultaine zwitterionic co-surfactant molecules. They showed resilient colloidal stability in seawater at elevated temperatures (up to 100 °C) for several months and reduced the seawater-crude oil interfacial tension (IFT) by three orders of magnitude. Thus, nano-transformation of the seawater-insoluble, yet highly efficient, abundant, and inexpensive surfactants enabled their utilization in oil recovery applications in high salinity and temperature oil reservoirs. Magnetic Nano-Mappers are coated magnetic nanoparticles that are colloidally and chemically stable under reservoir conditions. They enhance the magnetic permeability contrast between water and oil, thus allow in-situ monitoring of the oil-water interface using EM surveillance for optimal waterflooding operations.

537338 Invited Speaker: Polypeptoid Amphiphiles Serve As a Connective Glue to Build Lipid Layers on Vesicles: Fundamentals and Applications to Drug Delivery

Abstract:

Our recent work is based on the hydrophobic modification of biopolymers to provide building blocks for supramolecular architectures. The attachment of long chain alkyl groups to the polymer backbone allows the polymer to stick to hydrophobic surfaces and to â€œhookâ€ onto lipid bilayers. The hydrophobic effect exhibited by such systems can be used to design new functional nanostructures. For example, the use of such polymers to capture and tether liposomes leads to surfaces with densely packed liposomal layers that exhibit very low coefficients of friction in sliding lubrication, representative of articular joints. We also extend this concept to a system of hydrophobically modified polypeptoids (HMPs) which are amphiphilic pseudo-peptidic macromolecules with hydrophobic groups attached randomly along the polypeptoid backbone. We show that these biocompatible polymers connect across lipid bilayers and thus form layered structures on liposomes. The transition from a single bilayer to multiple bilayer structures is characterized by small angle neutron scattering (SANS) and cryo-transmission electron microscopy (cryo-TEM). Of specific interest is the observation that small bilayer rafts reattach to fresh unilamellar liposomes and self-assemble to form new two-bilayered liposomes reminiscent of two-bilayered organelles such as the nucleus in eukaryotic cells. These observations have significance to designing new nanoscale drug delivery carriers. Initial results on the *in-vitro* delivery of a highly hydrophobic chemotherapeutic agent, sorafenib, to hepatocellular carcinoma cells will be described.

539462 Tunable Supramolecular Assembly of Nucleoside Phosphoramidate Nanofibres By Enzyme Activation

Abstract:

The non-covalent association of molecules is a ubiquitous feature of life. Protein scaffolds, components of the extracellular matrix, and even duplexed nucleic acids all undergo supramolecular assembly. To generate synthetic self-assembled nanostructures, a wide variety of small molecule motifs have been developed that utilize hydrogen bonding, electrostatics, and hydrophobic interactions to contribute to the assembly process. Growing interest in the regulation of supramolecular assembly has produced a variety of responsiveness motifs. The governing effect of these motifs is exerted through physical and chemical cues such as ultrasound, temperature, pH, reduction and oxidation, and enzymatic responsiveness. Enzymes are ideal for the regulation of supramolecular interactions as they allow the mimicry of biological systems. They are catalytic in nature, able to be regulated by various biomolecules and ligands, and are biocompatible.

To create a responsive and tunable self-assembly system, we have developed Histidine Triad Nucleotide Binding Protein 1 (HINT1) responsive nucleoside phosphoramidate pro-gelators (PPGs). HINT1 has been well characterized as a nucleoside phosphoramidase and acyl-adenylate hydrolase. At PPG concentrations above the critical micelle concentration and below the gelation point, these molecules assemble into highly regular nanofibers resulting in bulk viscous liquid formation. Utilizing HINT1, the self-assembling peptides may be released from the blocking effect of nucleoside phosphoramidate moieties which induces the soluble nanofibers to condense into highly associated nanofiber bundles observed by electron microscopy. The structural transition to nanofiber crosslinking at the nanoscale results in bulk material gelation. We have utilized chemical biological tools in conjunction with small amplitude oscillatory rheometry to further characterize the role of HINT1 in the observed gelation event. Small molecule inhibitors and catalytically dead HINT1 mutants were used to investigate the role of HINT1 active site and catalytic activity on the regulation of PPG assembly. Inhibitors have been shown to block HINT1 activity on PPG substrates, and Hint1 H112N catalytically dead mutant has been shown to be unable to activate self-assembly. Our goal is to develop an adaptable system for the construction of biologically responsive materials that may be assembled *insitu* in response to HINT1 activity. Additional studies reporting the functionalization of these unique biomaterials will also be reported.

527072 Synergistic Assembly of Peptide Amphiphiles for Encapsulation of Camptothecin

Abstract:

Peptide amphiphiles(PAs) are a versatile class of self-assembling materials consisting of hydrophobic and hydrophilic segments which form supramolecular nanostructures like fibers, micelles and tubes. Self-assembly process involves a diverse range of forces like electrostatic interaction, hydrogen bonding, hydrophobic interaction and van der Waals force. The balance between these forces is dictated by molecular design to fabricate ordered structures with a high degree of atomic precision, regulated morphology and size. The hydrophobic region of the PAs form a hydrophobic core and the hydrophilic region projects outward on the surface of PA. This hydrophobic core can be exploited as a carrier for hydrophobic drugs.

In this study, camptothecin is chosen as the model hydrophobic drug for encapsulation. Currently, it is being used in cancer chemotherapy in the form of its analogues for the treatment of lung, colon and prostate cancer. However, its analogues impose certain clinical limitations as they cause an inactivation of its lactone form in blood and require prolonged infusions to prevent reversal of DNA cleavage. These limitations can be overcome by encapsulation of camptothecin in PAs by a co-assembly method. Two oppositely charged PAs sequences mutually screen the charges of each other in a synergistic manner to form hybrid drug encapsulated nanofibers. This novel method of encapsulation negates the use of external factors like pH, temperature and ions to bring about the self- assembly process and can also be used to design bifunctional nanofibers. Studies reveal that camptothecin was effectively encapsulated into the PAs fibers by a combination of hydrophobic and electrostatic interaction. Circular dichroism spectroscopy studies on the PAs-drug complexes demonstrate that the PAs retain their secondary structure and functionality when bound to the drug. The drug camptothecin was also found to exist in its active lactone form after encapsulation inside the PAs. *In vitro* drug release and cell studies were also performed to evaluate the controlled release and antitumor activity of camptothecin encapsulated in PAs. Overall, this PAs-drug system improves the bioavailability and cellular uptake of hydrophobic drugs in an aqueous environment due to their high aspect ratio.

535162 Structural Evaluation of Designer Co-Assembling Peptide Nanofibers

Abstract:

Co-assembling peptides represent an emerging platform for preparing supramolecular biomaterials with desired structures and functions for various medical and biotechnology applications. Compared to self-assembly where peptide interactions with itself dictate assembly, co-assembly only occurs upon interaction between complementary peptide partners. Recently, two designer Î²-sheet-forming co-assembling pairs, CATCH(+)/CATCH(-) peptides developed by Seroski and Hudalla and KW(+)/KW(-) peptides designed by King and Webb, have been reported. Key to these co-assembling peptide designs is the principle of charge complementarity where a net positive or negative charge on each peptide molecule discourages self-assembly, but electrostatic interactions between oppositely-charged peptide strands allow for co-assembly. Prior biophysical measurements using Thioflavin T fluorimetry, circular dichroism, and FTIR spectroscopy suggest both peptide systems co-assemble into Î²-sheet peptide nanofibers. Here, we examine at a molecular-level the role of charge on nanofiber structure using coarse-grained molecular dynamics simulations and solid-state NMR measurements. Both experimental measurements and computational predictions support the hypothesis of molecular-level co-assembly with Î²-sheets predominantly comprised of alternating peptide strands. Surprisingly, coarse-grained MD simulations and dipolar recoupling NMR measurements indicate peptides with the same charge can be adjacent to each other to varying degrees. Through a variation in net peptide charge, we demonstrate the possibility to control the distribution of nearest neighbors within co-assembled Î²-sheet peptide nanofibers.

535207 Evidence for Self-Assembly-Driven *Trans*-to-*Cis* Amide Bond Isomerization in Peptoid Nanosheets

Abstract:

Peptoids are synthetic polymers developed with the goal of mimicking the folding and functions of peptides. Peptoids have the same backbone structure as peptides, but with side chains that branch from the amide nitrogen instead of the Î ±-carbon. For biomaterials applications, peptoids benefit from the ability to adopt peptide-mimetic 3-dimensional structures but with decreased susceptibility to enzymatic degradation. Here we focus on peptoid B28 which adopts a Î £-strand conformation, analogous to Î²-strands in peptides, and self-assembles into nanosheets. Potential nanosheet applications have been proposed in the areas of catalysis, membranes, sensing, and molecular recognition. However, structural studies have proven difficult. To fill this knowledge gap, we have applied ^{13}C-^{13}C dipolar recoupling solid-state NMR measurements to reveal the configuration of backbone amide bonds selected by ^{13}C isotopic labeling of adjacent Î±-carbons. Measurements on the same molecules in the amorphous and assembled states revealed that amide bonds in the center of the amino block of peptoid B28 favor the *trans* configuration in the amorphous state and the *cis* configuration in the nanosheet. This unexpected result contrasts with previous NMR and theoretical studies of short solvated peptoids. Furthermore, examination of the amide bond at the junction of the two charged blocks within B28 revealed a mixture of both *cis* and *trans* configurational states, consistent with a previously-predicted brickwork-like intermolecular organization. The data we show here indicate a self-assembly driven isomerization process. This is a previously unreported phenomenon in peptoid biology and should be a point of consideration for peptoid self-assembly studies moving forward.

528196 Redox Sensitive Protein Droplets from Recombinant Oleosin

Abstract:

Molecular biology enables the creation of novel biomaterials with designer functionality and tailored responsiveness. In this work, we designed a protein that at low concentrations forms nanometer sized spherical micelles, and at high concentrations forms micron sized liquid droplets. Our work is based on the plant surfactant protein, oleosin. Oleosin, akin to a triblock copolymer, contains N and C-terminal hydrophilic regions and a central hydrophobic region. At low concentrations, the hydrophobic regions drives the formation of spherical micelles. Whereas, at higher protein concentrations, oleosin condenses into liquid droplets below a critical temperature. These liquid droplets are dynamic structures that fuse upon contact with one another and whose phase behavior is thermo-reversible, similar to many naturally occurring membrane-less organelles. In this work we explore strategies to control the phase behavior by introducing a cysteine residue to oleosin, a protein that does not contain any cysteines. Adding a cysteine residue to the N-terminal hydrophilic region of oleosin results in phase separation at a lower protein concentration and an increase the phase transition temperature. Furthermore, adding a reducing agent, beta-mercaptoethanol, to phase-separated, cysteine-containing oleosin rapidly dissolves the droplets. We demonstrate two strategies to precisely control the transition temperature of cysteine-containing oleosin. First, the transition temperature is tuned by varying the location of the cysteine in the oleosin backbone. Second, the transition temperature is tuned by blending the parent cysteine-less molecule with the cysteine containing mutant. These methods provide novel ways to control protein droplet formation and dissolution. This work will have applications as a system for the release of molecules with sensitivity to reducing conditions and as an engineered mimic of membrane-less organelles in synthetic protocells.

534032 A Way for Controlling Levan Nanoparticles Production Andparticle Size Distribution

Abstract:

Levan nanoparticles are produced via the self-assembly of a polyfructose biopolymer produced by some specific microorganisms.
However, the use of microorganisms for biopolymer production requires a deep control of many variables such as culture medium agitation, speed, oxygenâ€¦Therefore, cell-free systems, which use the isolated enzyme, have been proposed as a feasible alternative to overcome these drawbacks and to improve the uniformity of polyfructose in terms of molecular weight, and therefore of the levan nanoparticles formed by the polyfructose self-assembly.
In this work polyfructose is prepared by the cell-free polymerization reaction and the effect of the different sugar monomers has on the particle size, polymer molecular weight and the polymer production rate is described.
Results show that particle size can be modified (from 150 nm to 250 nm) by controlling the amount of sucrose and lactose in the culture medium. Furthermore, the addition of different sugars modifies the branching and therefore the polymer properties.
Finally, our results indicate that the addition of different sugars would permit the synthesis of levan nanoparticles à-la carte depending on the future use.

529739 Electronic Structure of Electron-Irradiated Graphene and Effects of Hydrogen Passivation

Abstract:

Defect engineering through irradiation processes and chemical functionalization of graphene are promising routes for fabrication of carbon nanostructures and two-dimensional metamaterials with unique properties and function. In previous computational studies, we reproduced experimentally observed structures of electron-irradiated graphene sheets through introduction of random distributions of vacancies in the graphene lattice and proper structural relaxation. We found that a vacancy-induced crystalline-to-amorphous transition in graphene occurs for an inserted vacancy concentration between 5% and 10%. This amorphization transition is accompanied by changes in grapheneâ€™s thermomechanical response, including a brittle-to-ductile transition in the fracture mechanism upon uniaxial tensile straining as well as a transition in the lattice thermal transport mechanism in these irradiated graphene sheets.

Here, based on molecular-dynamics (MD) simulations in conjunction with first-principles density functional theory (DFT) calculations, we report results for the electronic structure of irradiated and irradiation-induced amorphized graphene. We find that localized states appear at the Fermi level upon irradiation and the corresponding local density of states increases with increasing inserted vacancy concentration. Furthermore, electronic band structure calculations show that band flattening occurs due to electron localization in the vicinity of irradiation-induced defects and reduces the charge carrier mobility. This band flattening effect becomes stronger with increasing vacancy concentration inducing an increasing number of flat bands near the Fermi level. Moreover, we present electron wave functions (as frontier orbitals) and local charge density distributions, which provide clear evidence of carrier localization near the irradiation-induced carbon dangling bonds. Passivating these bonds with hydrogen atoms leads to delocalization of the charge density, hence increasing the carrier mobility, which also is seen in the reduced density of states observed at the Fermi level and the increased band dispersion with increasing inserted vacancy concentration. Importantly, we find these spatially localized states to be spin polarized, which gives rise to a net local magnetic moment. Passivation of these states can cause the complete removal of these induced local magnetic moments. Our studies set the stage for understanding and designing electronic two-dimensional materials for specific applications using irradiated graphene and passivated irradiated graphene as a well understood template.

529877 Development of a Novel Nanosensor Platform By Noncovalent Surface Engineering of Two-Dimensional Graphene Quantum Dots

Abstract:

Fluorescent nanomaterials have formed the core of nanosensor technologies with optical readouts, for real-time study of spatiotemporal dynamics in complex biological systems. In particular, electrostatic pinning of molecular recognition elements and biomimetic polymers on the surface of single-walled carbon nanotubes (SWCNTs) enables *selective recognition* of target biomolecules by the polymer and *signal transduction* by the near-infrared-emissive SWCNT [1]. In this work, we expand our nanosensor signal transduction element from a one-dimensional carbon nanotube to a two-dimensional graphene quantum dot (GQD). The flat surface of the GQD, in comparison to the highly curved SWCNT surface, provides an orthogonal opportunity to adsorb planar, sheet-forming polymers and thus allows further versatility in nanosensor design.

Existing work has focused on creating graphene sheet-based nanosensors and/or relies on covalent surface modification [2]–[4]. Here, we aim to create a new basis of fluorescent nanosensor development by (i) synthesizing GQDs with nanoscale dimensions that are accordingly the same order of magnitude in size as the biological systems they are designed to probe and (ii) employing noncovalent surface attachment to preserve the pristine graphitic domains that are responsible for the fluorescence emission properties. Although theory has predicted the ability of DNA nucleobases to adsorb to graphene surfaces [5]–[7] and this has been shown experimentally [8], [9], there has not yet been experimental confirmation for this system of noncovalently attached biopolymer to GQDs. We present a GQD-based sensing platform with accompanying comparisons of various GQD synthesis methods and GQD-polymer conjugate characterization. We first compare GQD synthesis schemes to create GQD preparations of varying sizes and oxidation levels. The GQDs are subsequently functionalized with a library of different polymers, including single-stranded DNA (ssDNA), synthetic nanosheet-forming peptoids [10], and phospholipids. Polymer adsorption is confirmed by a modulation in the fluorescence spectrum as well as AFM imaging. The role of GQD oxidation level, as determined by XPS analysis, and solution ionic strength are shown to play critical roles in the ability of polymers to adsorb to the GQD surface. GQDs with a lower oxidation level contain a larger proportion of` graphitic, sp^2-conjugated carbon domains, and thus enable stronger binding to polymer coatings via pi-pi stacking interactions between aromatic rings. Comparisons with prior theoretical work on polymer-GQD molecular interactions are discussed. In summary, the development and characterization of this two-dimensional GQD sensing platform will enable a broader range of adsorbed polymers and thus expand sensing capabilities.

References

[1] J. Zhang *et al.*, "Molecular recognition using corona phase complexes made of synthetic polymers adsorbed on carbon nanotubes," *Nature Nanotechnology*, vol. 8, no. 12, pp. 959–968, Nov. 2013.

[2] S. Zhang *et al.*, "A label-free and universal platform for the construction of an odd/even detector for decimal numbers based on graphene oxide and DNA-stabilized silver nanoclusters," *Nanoscale*, vol. 9, no. 33, pp. 11912–11919, 2017.

[3] H. Zhang *et al.*, "Universal fluorescence biosensor platform based on graphene quantum dots and pyrene-functionalized molecular beacons for detection of microRNAs," *ACS applied materials & interfaces*, vol. 7, no. 30, pp. 16152–16156, 2015.

[4] Y. Wang, L. Tang, Z. Li, Y. Lin, and J. Li, "In situ simultaneous monitoring of ATP and GTP using a graphene oxide nanosheet–based sensing platform in living cells," *Nat. Protocols*, vol. 9, no. 8, pp. 1944–1955, Aug. 2014.

[5] H. S. Kim, B. L. Farmer, and Y. G. Yingling, "Effect of Graphene Oxidation Rate on Adsorption of Poly-Thymine Single Stranded DNA," *Adv. Mater. Interfaces*, vol. 4, no. 8, p. n/a-n/a, Apr. 2017.

[6] J. Antony and S. Grimme, "Structures and interaction energies of stacked graphene–nucleobase complexes," *Phys. Chem. Chem. Phys.*, vol. 10, no. 19, pp. 2722–2729, May 2008.

[7] J.-H. Lee, Y.-K. Choi, H.-J. Kim, R. H. Scheicher, and J.-H. Cho, "Physisorption of DNA Nucleobases on h-BN and Graphene: vdW-Corrected DFT Calculations," *J. Phys. Chem. C*, vol. 117, no. 26, pp. 13435–13441, Jul. 2013.

[8] N. Varghese *et al.*, "Binding of DNA nucleobases and nucleosides with graphene," *ChemPhysChem*, vol. 10, no. 1, pp. 206–210, 2009.

[9] S. Manohar, A. R. Mantz, K. E. Bancroft, C.-Y. Hui, A. Jagota, and D. V. Vezenov, "Peeling Single-Stranded DNA from Graphite Surface to Determine Oligonucleotide Binding Energy by Force Spectroscopy," *Nano Letters*, vol. 8, no. 12, pp. 4365–4372, Dec. 2008.

[10] B. Sanii *et al.*, "Shaken, not stirred: Collapsing a peptoid monolayer to produce free-floating, stable nanosheets," *Journal of the American Chemical Society*, vol. 133, no. 51, pp. 20808–20815, 2011.

530126 Interpretation of the Far-Infrared Optical Spectrum of SWCNTs and Graphene

Abstract:

Both metallic and semiconducting SWCNTâ€™s have a broad optical absorption peak that spans the terahertz and far-infrared regions of the electromagnetic spectrum. This feature is, for the most part, absent in graphene.

This talk will review new and existing experimental evidence for the assignment of this unique optical peak found only in SWCNTs (â€œrolled upâ€ graphene) but not planar graphene itself. Once we motivated the assignment of this peak we will consider how the far-infrared region of the electromagnetic spectrum can be utilized to characterize the doping and/or functionalization, the degree of dispersion and alignment of SWCNTs and SWCNT films.

530996 Formation and Thermomechanical Behavior of Nanocomposite Superstructures from Interlayer Bonding in Twisted Bilayer Graphene

Abstract:

Graphene-based nanomaterials have exceptional electronic, mechanical, and thermal properties that can be tuned by precise control of their nanostructural features. Such tunable properties are responsible for the unique function of these nanomaterials that has potential to enable numerous technological applications. Toward this end, we report here a systematic computational analysis of the formation and mechanical behavior of a class of two-dimensional (2D) carbon-based nanostructures, namely, graphene-diamond nanocomposites formed through interlayer covalent bonding of twisted bilayer graphene with commensurate bilayers. The interlayer bonding is induced by patterned hydrogenation that leads to formation of superlattices of 2D nanodiamond domains embedded between the two graphene layers with the periodicity of the underlying Moiré pattern. We have conducted a comprehensive classification of all such superstructures that can be formed with respect to the corresponding structural parameters, including the commensurate bilayerâ€™s twist angle, the stacking type of the nanodomains where the interlayer bonds are formed, the interlayer bond pattern and density, and the concentration of sp^3-bonded C atoms in these superstructures.

We also report results of a systematic study of the mechanical behavior of such carbon nanocomposite superstructures based on molecular-dynamics (MD) simulations of uniaxial straining tests according to a reliable interatomic bond-order potential. The mechanical properties of these 2D materials are investigated over the full range of structural parameters. We introduce a ductility metric, demonstrate its direct dependence on the concentration of sp^3-bonded C atoms, and show that increasing the concentration of sp^3-bonded C atoms beyond a critical level induces ductile mechanical response. We analyze the ductile fracture mechanisms, mediated by void formation, growth, and coalescence, in contrast to the typical brittle fracture of graphene and probe the brittle-to-ductile transition. In addition, we report results of MD simulations of nanoindentation tests that we have performed on these superstructures, including mechanical properties of these 2D materials, such as elastic modulus, strength, and toughness, and demonstrate their direct dependence on the superstructuresâ€™ structural parameters as well as indentation parameters. The underlying fracture mechanism of such superstructures under nanoindentation testing also is characterized in detail. Finally, we report results of non-equilibrium MD simulations of thermal transport in these superstructures, aimed at determining the dependence of the lattice thermal conductivity of the superstructures on the key structural parameters, such as the concentration of covalent interlayer C bonds. Our study sets the stage for designing few-layer-graphene-based nanocomposites with unique thermomechanical properties and functionality over a broad range of applications.

536381 Electrochemically Triggered Nucleation and Growth of Zinc Phosphate Co-Deposited with Amino-Modified Graphene Oxide

Abstract:

Electrochemically assisted phosphating has been proposed and investigated for decades because of its energy efficient and eco-friendly. Herein, graphene oxide (GO) is first grafted with vinyl group through the reaction with Triethoxyvinylsilane. Then, the vinyl group modified GO is further reacted with [2-(Methacryloyloxy)ethyl]trimethylammonium chloride, incorporating amino group into GO sheets(AGO). Due to the incorporated amino groups, AGO sheets are positively charged in water and thus can be stably dispersed in the electrolyte that containing various ions under strong acid condition. After that, AGO is added to the electrolyte solution as modifier of the electrochemical phosphating process, and then co-deposited with phosphate coating on the steel substrate. The morphology and chemical composition of AGO modified phosphate coating are characterized via scanning electron microscopy and X-ray diffraction. Based on that, the possible mechanism of the AGO assisted electrochemical phosphating is proposed. Electrochemical measurements, including polarization curves and electrochemical impedance spectroscopy (EIS), are then used to evaluate the corrosion inhibitive properties of AGO modified phosphate coating.

524321 Dispersion Microstructure and Aerogel Properties of Graphene/Manganese Oxide Mixtures and Hybrids

Abstract:

The focus of this work is to investigate the effects of dispersion properties on the porous three-dimensional microstructure and electrochemical properties of aerogels produced by freeze-drying ultra-large reduced graphene oxide (RGO) and manganese dioxide nanowire (MnO_2) dispersions. The inclusion of MnO_2 nanowires as guest materials in assemblies of ultra-large RGO was achieved via two techniques: a) mixing both dispersions that have been synthesized separately to form a mixture and b) growing the nanowires along the ultra-large graphene sheets to form a hybrid material. Scanning electron and atomic force microscopy were used to image the starting materials and gain initial insights into the interactions between the nanowires and sheets. The effects of concentration on the microstructures of dispersions of these materials were investigated using polarized optical microscopy (POM) and rheology. Next, aerogels were produced by freeze-drying hybrids and mixtures with different concentrations and qualitatively assessed for mechanical integrity. Aerogels that could retain their shape with handling were further characterized for morphology using SEM, specific surface area using Brunauerâ€"Emmettâ€"Teller (BET) and capacitive behavior using cyclic voltammetry. Further understanding the relationships between the initial materials, their dispersion microstructure, and the resulting aerogels will enable their application for capacitive deionization and other electrochemical applications requiring high electrical conductivity, large surface area, and interconnected porous architectures.

539865 Invited Talk: Towards In Vivo Bioimaging of Electrical Fields and Mechanical Forces with Stimuli-Responsive Upconverting Nanoparticles

Abstract:

Cellular signaling and health are governed in large part by mechanical forces and electromagnetic fields. Mechanical forces play a critical role in cell differentiation, tissue organization, and diseases such as cancer and heart disease, whereas membrane voltage transients (i.e., action potentials) are essential for inter-cellular signaling, muscle contraction, and sensory perception. Accordingly, quantifying a biological systemâ€™s forces and fields is crucial for understanding physiology and disease pathology, as well as for developing medical tools for repair and recovery. Here, we introduce a new class of *in vivo*probes to monitor biological forces and fields at the organ-to-organism scale with high spatial and temporal resolution. Our design is based on inorganic upconverting nanoparticles that, when excited in the near-infrared, emit light of a different color (wavelength) and intensity in response to sub-100mV electric potentials and nanoNewton-to-microNewton forces. The nanoparticles are sub-20nm in size, do not bleach or photoblink, need not be genetically encoded, and, by virtue of their infrared absorption, can enable deep tissue imaging with minimal tissue autofluorescence and light-induced tissue damage. We present the design, synthesis, and characterization of these nanoparticles both in vitro and in vivo, focusing on the forces generated by the roundworm *C. elegans*as it feeds and digests its bacterial food. Chronic and acute cytotoxicity assays are also used to confirm biocompatibility. Our force measurements are coupled with electrical measurements of muscle contractions in both wild-type and mutant animals, providing insight into the interplay between mechanical, electrical, and chemical signaling *in vivo*.

537832 Development of Lspr-Based Biosensor for the Detection of Sjogrenâ€™s Syndrome Biomarkers

Abstract:

Sjogrenâ€™s syndrome (SS) is a chronic autoimmune disease that commonly manifests through oral and ocular dryness as a result of lymphocytic infiltration of the salivary and lacrimal glands. Patients with SS often go undiagnosed due to the lack of specific symptoms. Recently, there has been growing interest in using biomarkers in the saliva and lacrimal fluid to develop non-invasive tests for SS that can provide rapid and accurate feedback. Despite the promise that these methods hold, many potential biomarkers lack the specificity required, and patients with reduced lacrimal and salivary function may have difficulty producing enough fluid for quantitative measurement. For this reason, this work looks at the design and evaluation of utilizing the local refractive index sensitivity of the localized surface plasmon resonance (LSPR) of gold nanoshells (AuNS) for the detection of protein biomarkers expressed in patients with SS.

AuNS were prepared by seeded growth of colloidal gold on aminated silica nanoparticles and encapsulated by poly(maleic anhydride-*alt*-1-octadecene)-g-poly(ethylene glycol) methacrylate (PMAO-g-PEGMA) graft copolymer to improve particle stability. Transmission electron microscopy (TEM) was used to determine the size distribution. Then, the AuNS were incorporated into the aqueous phase of an inverse emulsion copolymerization of acrylamide and methacrylic acid, crosslinked with poly(ethylene glycol) dimethacrylate (PEGDMA), to form a polymer shell on the surface of AuNS that can be further modified by altering the composition and the PEGDMA chain length to affect the swelling of the polymer network. Additional comonomers with desirable functional groups can be introduced in order to tune the affinity for the target protein. For optimization of the sensor design, the impact of different substrate anchoring methods for the AuNS will be investigated. The AuNS density and the sensitivity of the LSPR response will be determined using TEM and UV-VIS spectroscopy, respectively and compared to the response of the functionalized AuNS in solution.

Work supported by National Institutes of Health Grant R01-EB022025

537382 Development of Protein-Gold Nanoparticle Based Colorimetric Radiation Sensor

Abstract:

High energy X-ray radiation has been used for ablation of malignant tumors for quite some time. Nevertheless, medical practitioners are unable to predict with high accuracy the amount of radiation reaching the point of interest below the tissue level. To help address this issue, we are developing a colorimetric sensor for radiation using the gold nanoparticle forming property of bio-molecules (particularly Maltose Binding Protein (MBP)) in presence of ionizing radiation. As gold nanoparticles have distinct color, gold nanoparticles formed on irradiation changes the color of the solution and thus can be used as colorimetric sensor. The color change is observed for doses of X-ray radiation as low as 2 grays (1 gray (Gy) = absorbance of 1 joule of radiation energy per kilogram of matter). The system also shows a radiation dose dependent color shift (from 3 Gy to 15 Gy), which can be used to discriminate between different delivered doses.

534998 Characterizing the Uptake of Quinic Acid and Tannic Acid Coated Iron Oxide Nanoparticles for Labeling of Cancer Cells

Abstract:

Brain related malignances are extremely aggressive and current diagnostic as well as therapeutic techniques have proved largely ineffective in targeting these tumors. In the context of diagnostics, methods to label cancer cells during and after the onset of carcinogenesis are being investigated. Herein, we have investigated the utility of ultrasmall iron oxide nanoparticles (USIONPs) to label U87, glioblastoma cell line (primary brain cancer) and MDA-MB-231Br, brain metastasizing variant of MDA-MB-231 triple negative breast cancer cells. Specifically, we have characterized the uptake of two types of USIONPs coated with quinic acid and tannic acid by U87 and MDA-MB-231Br cells. Based on the Prussian blue staining (qualitative) and ferrozine colorimetric iron detection (quantitative), we determined that the uptake of quinic acid coated USIONPs was significantly greater compared to that of the tannic acid coated USIONPs. In the case of MDA-MB-231Br cells, the uptake of quinic acid coated USIONPs was ~3 fold higher than the uptake of tannic acid coated USIONPs. Further, it was observed that the ratio of the uptake of quinic acid coated USIONPs to the tannic acid counterpart was highest at the early time point (i.e.,4 hr) and decreased at later time points (i.e., 72 hr). Endocytosis mainly mediated the uptake of USIONPs however, increased uptake of quinic acid coated USIONPs at earlier time point suggests that they have a tendency to selectively bind to the cancer cells before being up taken whereas, the tannic acid coated USIONPs are non-selectively up taken through endocytosis. In addition, we have observed that both versions of USIONPs do not significantly impact the proliferation of either cell line. Thus, these cells can be effectively labeled without compromising cell viability. We are currently performing transmission electron microscopy to confirm that quinic acid coated USIONPs are indeed selectively binding the cancer cells before being up taken. We are also investigating the differences between the uptake of quinic acid versus tannic acid coated USIONPs; specifically, the role of selectins in mediating the selective uptake of quinic acid coated USIONPs. This mechanistic understanding can lead to more effective targeting of cancer cells for diagnostic and therapeutic purposes.

531005 Tuneable Mechanical Response of Twisted DNA Nanotubes Towards Biosensing

Abstract:

Tuneable mechanical response of twisted DNA nanotubes towards biosensing
S. Kumar and R.E. Taylor

Helical motifs can be found abundantly in the biological world of micro- and nanostructures such as microbial flagella and cellulose fibrils. As engineering components, helical motifs exist in the macroscopic world as screws and springs. Downsizing these simple tools into the micron scale would help towards understanding mechanical forces and motion that regulate the function of numerous mechanosensory systems. Examples of these systems include motor proteins and cytoskeletal filaments. Multi-helical DNA bundles could serve as a model nanomaterial to build nanoarchitectures that could sense or mimic such biological molecules and systems.

We focus here on the reversible conformational change of tuneable twist features on DNA nanotubes. The DNA nanotubes are self-assembled into their pre-stressed twisted state through specific and localized basepair insertions on select DNA single-stranded tiles (SST). Using microscopic techniques including TEM and TIRF microscopy, we show that that these twisted DNA nanotubes can reversibly twist and untwist via toe-hold mediated strand displacement of incoming fuel DNA strands. Our results also show the capability of these systems to undergo reversible twisting when substrate- or interface-bound. These results highlight the capability of these twisted DNA nanotubes to be used as mechanical biosensors in biological systems and tuneable microswimmers in biomimetic applications.

521840 Novel Fluorescent Nano Structures for Bio-Imaging of MCF-7 Cells

Abstract:

Novel Fluorescent Nano Structures for Bioimaging of MCF-7 Cells

Aishee Dey*, Lopamudra Giri# and Sudarsan Neogi*

* Indian Institute of Technology, Kharagpur

Indian Institute of Technology, Hyderabad

Although nanomaterials have a wide area of application including targeted drug delivery, diagnostic imaging and controlled release, the design of nanomaterials having multimodal applications remains challenging. The recent trend in the field of nanomedicine revolves around development of in vivo dignostics and therapeutics that can be used as theranostics. In order to achieve this, we performed the optimization of the fabrication process at multiple levels.The major objective of the current work is to tune unique physicochemical properties in nanostructures through proper selection of fabrication methods, that can be used as therapeutic as well as diagnostic tools. Here we choose to develop a fluorescent imaging sensor based on metal oxide nanoparticles that can be used to study the nanoparticle uptake in cells and penetration of nanoparticles through tissue through laser scanning confocal microscopy [1]. Specifically, we choose to use semiconductor quantum dots or metal oxide nanoparticles as they can be designed to have optical, magnetic, and structural properties that are different from the same particles in their bulk form [2]. Here we choose to use ZnO nanoparticles as they are known to provide a targeted delivery which is beneficial for cancer patients as they can differentiate between cancer cell and normal cells, and hence avoid their adverse effects on normal cells.Secondly, we chose to use ZnO nanoparticles because of their excellent bio-compatibility [3]. Another feature of the proposed work is that the targeted approach requires minimal drug loading in contrary to large drug loading required in case of non-targeted approach [4]. The major novelty of the work lies in achieving the multimodalities in a single nanostructure, where we manipulated the optical properties of ZnO so that it can be used as therapeutic reagent as well as imaging sensor to study the cell uptake and penetration of particles.

Here we propose a surfactant assisted method to synthsize the fluorescent ZnO nanoparticles. ZnO particles were formed using chemical synthesis method by mixing zinc nitrate hexahydrate and ammonium carbonate solutions along with a surfactant Tween 80 [5]. Since the objective of the work was to maximize the drug efficacy along with maintenance of the fluorescent level of the nanostructures, a series of experiment were performed for nanoparticle fabrication using various concentrations of Tween-80. The result clearly indicates that with increasing amount of Tween 80 present in the mixture, the morphology of the nanoparticles changed along with the fluorescence of the particles. The fluorescent nature of the particles increased with increase in Tween 80 which was evident from the confocal microscope imaging, the best being observed at Tween 80 concentration of 14% (v/v) (ZnOT14) . In the excitation wavelength of 488 nm and an emission range of 508-550 nm, the nanoparticles showed green fluorescence as shown in figure 1. The nanoparticles were further characterized by SEM, TEM, XRD and FTIR techniques. Images from the SEM and TEM show that with change in Tween 80 concentration, the size of the nanoparticles change as demonstrated in figure 2. XRD results confirm the formation of ZnO due to presence of prominent peaks at 31.9°, 34.6°, 36.61°, 47.88°, 68.34°, 69.1° which are in good agreement with JCPDS 89-0511. As per FTIR results, broad peak at 3500 cm^{-1} indicating the presence of hydroxyl groups intensified on addition of Tween 80, as the surfactant introduces additional hydroxyl groups on the surface of the nanoparticles. The ZnO nanoparticle with Tween 80 concentration of 14% (v/v) was found to be highly haemocompatible (0.031% haemolysis), indicating its non-toxicity. Secondly, the nanoparticles were used for imaging of MCF-7 cells which are a breast cancer cell line. Breast cancer is one of the most common cancer found in women. Breast cancer can lead to metastasis which is the secondary malignant growth from the primary cancer site. With progress in metastases of breast cancer, some organs like lungs, liver and bone get damaged [6]. Thus imaging of the cells by nanoparticles is important as it can lead to proper detection of the cancer tissues. Also the MCF-7 is an immortal cell line which do not proliferate indefinitely and can be grown in vitro. The cells were cultured in a media containing 44.5 ml Dulbecco's modified eagle's medium , 5 ml of 10% fetal bovine serum and 0.5 ml of 1% penstrap antibody in a 50 ml media. The cells were cultured for 4 days. Further the cells were treated with the synthesized ZnO nanoparticles and imaged using confocal microscope.

Figure 1: Variation of fluorescence of ZnO nanoparticles with Tween 80 concentration of (a) 0 % (v/v), (b) 6 % (v/v), (c) 12 %(v/v) and (d) 14 %(v/v).

b

a

d

c

Figure 2: FESEM images of ZnO nanoparticles with Tween 80 concentration of (a) 0 % (v/v), (b) 6 % (v/v), (c) 12 %(v/v) and (d) 14 %(v/v).

(b)

(a)

Figure 3: (a) Histogram showing percentage haemolysis of positive control, negative control and ZnOT14; (b) Original images of haemolysis experiment showing ZnOT14, positive control and negative contol.

References

[1] T. Xia *et al.*, "Comparison of the Mechanism of Toxicity of Zinc Oxide and Cerium oxide Nanoparticles Based on Dissolution and Oxidative Stress Properties, " *ACS Nano,* vol. 2, no. 10, pp. 2121-2134, 2008.

[2] D. Groneberg, G. Michael, T. Welte and U. Pison, "Nanoparticle-based diagnosis and therapy .,"*Curr Drug Targets,* vol. 7, no. 6, pp. 643-648, 2006.

[3] J. W. Rasmussen, E. Martinez, P. Louka, and D. G. Wingett, "Zinc Oxide Nanoparticles for Selective Destruction of Tumor Cells and Potential for Drug Delivery Applications ", *Expert Opin Drug Deliv*., vol. 7, no. 9, pp 1063–1077, 2010.

[4] S. Nie, Y. Xing, G. J. Kim, and J. W. Simons, "Nanotechnology Applications in Cancer," *Annu. Rev. Biomed. Eng.*, vol. 9, no. 1, pp. 257–288, 2007.

[5] Y. Khan, S. K. Durrani, M. Mehmood, J. Ahmad, M. R. Khan, and S. Firdous, "Low temperature synthesis of fluorescent ZnO nanoparticles," *Appl. Surf. Sci.*, vol. 257, no. 5, pp. 1756–1761, 2010.

[6] E. Goldman *et al.*, "Nanoparticles target early-stage breast cancer metastasis *in vivo*," *Nanotechnology*, vol. 28, no. 43, pp. 43LT01, 2017.

534019 Engineered Green Fluorescent Proteins: Cartilage-Targeted Delivery Nanocarriers That Provide Insights on the Effects of Charge on Transport into Dense Charged Tissues

Abstract:

Post-traumatic Osteoarthritis (PTOA) is a degenerative disease that affects cartilage and associated joint tissues. PTOA is triggered by a traumatic joint injury and there are currently no disease modifying osteoarthritis drugs to slow down or halt disease progression. A major challenge for developing such a therapeutic is associated with drug delivery into cartilage. Systemic delivery is ineffective because cartilage does not have a blood supply. If drugs are injected directly into the joint (intra-articular therapy), they are rapidly cleared by the synovium leading to poor cartilage penetration. Cationic delivery carriers such as Avidin have significantly improved the uptake, penetration and retention of low molecular weight drugs through electrostatic interactions with negatively charged aggrecan proteoglycans in cartilage.

We have characterized the precise effect of carrier charge on delivery efficacy into cartilage in isolation of other factors such as size and shape that can influence transport. While higher cationic charge leads to enhanced transport due to higher Donnan partitioning at the cartilage-synovial fluid interface, subsequent electrostatic binding interactions between the nanocarrier and cartilage matrix can slow transport. We studied the use of supercharged green fluorescent proteins (S-GFPs) engineered to have net charge ranging from +9 to +36 at physiological pH while having nearly identical sizes and shapes. As controls, we used three engineered GFP variants having net neutral charge but different surface charge distributions. All the S-GFPs had higher quantitative cartilage uptake compared to the controls. Amongst the S-GFPs, the lower charged variants (net charge +9 and +15) had significantly higher net uptake and penetration into cartilage compared to higher charged variants (net charge +25 and +36). These trends were observed both in bovine cartilage and in human cartilage obtained from two donors with no or early osteoarthritic cartilage. Two of the three neutral variants did not show preferential uptake into cartilage while the third variant had an uptake ratio (defined as the ratio of concentration inside the tissue to the final bath concentration) that was an order of magnitude higher than the other two. This demonstrated that, in addition to net cationic charge, surface charge distribution also plays a very important role in carrier transport in cartilage. Cellular uptake of S-GFPs into cartilage cells increased monotonically with higher cationic charge, which was the opposite of the trends observed in tissue-level uptake and penetration experiments. Based on these results, +9 GFP was found to be the optimal delivery carrier for drugs with ECM or cell surface receptor targets. For drugs with intracellular targets, a variant with intermediate charge, such as the +15 GFP, would provide a better balance between tissue penetration and cellular uptake.

536919 Chemically Tuned NIR Light-Activated Bionanoconjugates for the Selective Destruction of Tumor Cells in Heterocellular 3D Models and for Quantitative *In Vivo* tumor Imaging

Abstract:

The synthetic complexity involved in the nanofabrication of tumor specific bionanoconjugates and the lack of clinically relevant screening platforms to rapidly establish their selectivity are both major hurdles in their development. Using a fine-tuned synthetic chemical approach, we engineer NIR light-activated bionanoconjugates (NABs) directed towards the tumor-associated biomarker, epidermal growth factor receptor (EGFR), using the approved antibody cetuximab. In order to rapidly asses their efficacy, we developed a high-throughput 3D heterocelluar model of pancreatic ductal adenocarcinoma including patient-derived cancer-associated fibroblasts as a robust screening platform to evaluate the developed NABs. Using this 3D platform, we demonstrate that the NABs efficiently permeate the 3D tumor tissue within 1 hour and provide up to 8-fold binding specificity towards 3D tumor tissue. This resulted in a striking 16-fold improvement in targeted phototoxicity of the heterocellular 3D nodules following irradiation with non-toxic NIR light. To validate the tumor specificity of the NABs *in vivo*, we leveraged a powerful paired-nanotracer molecular imaging approach to validate their successful engagement with EGFR-positive tumors *in vivo* and are equally as available for binding as the natural ligand, EGF. The stepwise approach described here provides a rational workflow whereby the fine chemical tuning of NABs, combined with rapid testing in complex 3D tumor models and quantitative *in vivo* molecular imaging, promises to accelerate the translation of such therapeutic approaches with higher specificities and lower patient toxicities.

531033 Effects of Pore Morphology and Pore Edge Termination on the Mechanical Behavior of Graphene Nanomeshes

Abstract:

Graphene-based nanomaterials hold great promise for numerous technological applications because of their exceptional mechanical, electronic, and thermal properties. Graphene nanomeshes (GNMs) are ordered, defect-engineered graphene nanostructures consisting of periodic arrays of pores in the graphene lattice with neck widths less than 10 nm. The electronic, transport, and mechanical properties of GNMs can be tuned by varying the structural, chemical, and architectural parameters of the nanomeshes, such as their porosity as well as their pore lattice structure, pore morphology, and pore edge termination. Here, we study the mechanical response of GNMs to uniaxial tensile straining and determine their mechanical properties based on molecular-dynamics simulations of dynamic deformation tests according to a reliable bond-order interatomic potential. We establish the dependences of the fracture strain, ultimate tensile strength, and fracture toughness on the nanomesh porosity and derive scaling laws for GNM modulus-density and strength-density relations. We place special emphasis on establishing the dependence of the above properties on pore morphology for GNMs with circular and elliptical pores over a range of aspect ratios. In addition, we demonstrate elastic stiffening and strength reduction caused by pore edge termination with hydrogen atoms; these effects on the GNMsâ€™ mechanical behavior are directly related to the extent of hydrogenation, namely, strength and toughness are further reduced if the hybridization of the hydrogenated carbon atoms at the GNMâ€™s pore edge is changed from sp^2 to sp^3. The underlying mechanisms of crack initiation and propagation, and of nanomesh failure are characterized in detail and the effects of pore morphology and edge passivation on the fracture behavior are demonstrated.

532645 Influence of Sonication Conditions and Wrapping Type on Yield and Fluorescent Quality of Noncovalently Functionalized Single-Walled Carbon Nanotubes

Abstract:

As nanomaterials have become more accessible and manufacturable, nanoscale biosensors have become an increasingly relevant area of research. One such nanomaterial is the suspended single-walled carbon nanotube (SWNT), which fluoresces in the near-infrared window of biological material, making it ideal for in vivo applications. SWNT may be suspended in water if wrapped with an amphiphilic polymer or surfactant; the suspended SWNT can act as a simple optical probe or as a sensor by engineering the wrapping to have selective domains that cause a change in fluorescence when bound by a target analyte. The process of suspending nanotubes is typically achieved by sonication. While much application-focused research has been performed on suspended SWNT sensors and probes, little has been done to understand factors affecting SWNT fluorescent quality after suspension. We have explored effects of sonication power and duration on an array of 9 potential wrappings including proteins, DNA, oligosaccharides, polysaccharides, synthetic polymers, and surfactant solutions. Optimal sonication conditions were found to vary on an individual wrapping basis. Trends between polymer length and fluorescence properties suggest fragmentation of the wrapping to play a role in SWNT functionalization. These trends may be used to predict optimal processing conditions to create suspended SWNT with maximal start fluorescence for various wrapping types, improving sensor dynamic range and sensitivity. These results also point to the need for control of sonication conditions in large scale synthesis to ensure tighter batch to batch reproducibility of nanotube sensors.

537260 DNA-Controlled Brightening of Carbon Nanotube Photoluminescence in Acidic Environments

Abstract:

Single-wall carbon nanotubes (SWCNTs) are a family of one-dimensional (1D) nanostructures that have defined handedness and helicity. Each of the distinct structures of SWCNTs possess unique optical and electronic properties, which are sensitive to small perturbations in the nanotube environment. In this work, we investigated the dispersion stability and optical modulation of DNA-wrapped SWCNTs (DNA-SWCNTs) in aqueous environments for a broad pH range of 2.0-13.0. The stock dispersions of DNA-SWCNTs prepared by our standard preparation method showed near neutral pH values. Surprisingly, by adding hydrochloric acid, an increase in SWCNT fluorescence was observed in acidic environments with pH values around 3.0-5.0. The brightening of nanotube photoluminescence was examined to be strongly dependent on the DNA-sequence among other factors. These results will enable identifying stable DNA-SWCNT hybrids that are suitable for potential applications, including nanotube-based pH sensors.

535568 Ionic Strength-Mediated Phase Transitions of Surface-Adsorbed DNA on Single-Walled Carbon Nanotubes

Abstract:

Single-stranded DNA oligonucleotides have unique, and in some cases sequence-specific molecular interactions with the surface of carbon nanotubes that remain the subject of fundamental study. In this work, we observe and analyze a generic, ionic strength-mediated phase transition exhibited by over 25 distinct oligonucleotides adsorbed to single-walled carbon nanotubes (SWCNTs) in colloidal suspension. The phase transition occurs as monovalent salts are used to modify the ionic strength from 500 mM to 1 mM, causing a reversible reduction in the fluorescence quantum yield by as much as 90%. The phase transition is only observable by fluorescence quenching within a window of pH and in the presence of dissolved O_2, but occurs independently of this optical quenching. The negatively charged phosphate backbone increases (decreases) the DNA surface coverage on an areal basis at high (low) ionic strength, and is well described by a two-state equilibrium model. The resulting quantitative model is able to describe and link, for the first time, the observed changes in optical properties of DNA-wrapped SWCNTs with ionic strength, pH, adsorbed O_2, and ascorbic acid. Cytosine nucleobases are shown to alter the adhesion of the DNA to SWCNTs through direct protonation from solution, decreasing the driving force for this phase transition. We show that the phase transition also changes the observed SWCNT corona phase, modulating the recognition of riboflavin. These results provide insight into the unique molecular interactions between DNA and the SWCNT surface, and have implications for molecular sensing, assembly, and nanoparticle separations.

532747 Graphene Oxide Model Development Via Reactive Molecular Dynamics Simulations

Abstract:

The properties of graphene oxide (GO) are largely depending on the coexistence of hybridized carbon atoms, the diversity of oxygen-containing functional groups, and the structural defects. It is of vital importance to develop atomic graphene oxide model, which in turn enables the construction of structure-property relationship, as well as the design and upgrade of graphene oxide-based applications. Despite the efforts of graphene oxide model development over the last decade, there is no atomic GO model available, capturing all key characteristics of experimental GO samples, namely, the hybridization and distribution of carbon atoms, the C/O ratio, the size and distribution of vacancy defects. In this work, we carried out temperature-programmed reactive molecular dynamic simulations (RxMD) to generate atomistic graphene oxide models. By a systematical study of three controlling parameters: the initial functionalization density, the ratio of basal hydroxyl and epoxy groups, and the annealing trigger temperature, we propose an atomic GO model that is representative of the GO product via the modified Hummer method. The atomistic GO models have been analyzed and compared with available experimental results of Fourier-transform infrared spectroscopy (characteristic functional groups), X-ray photo-electron spectroscopy (hybridization and distribution of carbon atoms), and atomic force microscopy (thickness of curved GO). The new model provides insight specifically for GO membrane development and applications.

530803 Analysis of Surfactant Exchange Kinetics of DNA-Wrapped Carbon Nanotubes

Abstract:

Single-wall carbon nanotubes (SWCNTs) are one-dimensional nanostructures that have defined handedness and helicity. The unique optical and electronic properties of distinct SWCNT species offer possibilities in many promising technological applications. Particularly, the comprehensive SWCNT sorting by single-stranded DNA achieved in recent years has stimulated the understanding of structure-property relationships for structurally defined DNA-wrapped SWCNTs (DNA-SWCNTs). In this work, we measured the kinetics of surfactant exchange reactions of structurally sorted DNA-SWCNTs by monitoring spectral changes in the SWNCT absorption and fluorescence. The results were correlated to the binding strength of the specific SWCNT species and DNA recognition pair. In addition, the impact of environmental stimuli, such as temperature, on the dispersion stability and optical properties of DNA-SWCNTs has been investigated. These efforts will enable us to better understand the interactions of DNA-SWCNTs with certain environments and to design novel DNA-SWCNT hybrid materials for applications development.

535968 Invited Speaker: Real Time, Label Free Biosensing in-Vivo Using Single Walled Carbon Nanotubes and Other Carbon Nanomaterials

Abstract:

The human body is a vast network of biochemical signaling pathways dictating physiological and pathological processes. Our group is interested in developing nanometer-scale sensor technology that can probe this network for biomedical applications. We have previously introduced CoPhMoRe or corona phase molecular recognition as a method of engineering synthetic molecular recognition sites from a heteropolymer library. This technique has significant potential for using SWNT-based sensors to interface with biological systems, allowing monitoring pathways at the sub-cellular, cellular, tissue, and whole-animal scale. For this presentation, I will discuss several important points relating to nanoparticle characterization and the ability to generate effective sensors in-vivo. Techniques that can quantify the particle size distribution of nanoparticles rapidly and accurately will be shown as critical to optimizing signal and sensor response. For DNA-wrapped carbon nanotubes as a model nanoparticle sensor, we highlight phase transitions in the corona phase on the nanotube surface that are ionic strength driven. The ability for corona phase to exhibit such phase transitions at the nanotube surface has important implications and opportunities for sensors and molecular recognition. My talk will also cover our efforts at MIT to enable in-vivo sensors for biomedical applications. Animal experiments for rodent and marine animal model will be described with technological hurdles highlighted for realizing the real time, label free detection of analytes in-vivo using nanosensor technologies. Lastly, we will also highlight our efforts to create autonomous state machines for medical applications, including conceptual and modeling efforts, as well as prototype colloidal nanoparticles that can autonomously sense, electronically store and report information, bringing sensors and nanoelectronics to remote locations, including the human digestive tract and ultimately the circulator system.

538401 Graphene Based Sensing Platform for Studying Amyotrophic Lateral Sclerosis

Abstract:

Amyotrophic lateral sclerosis (ALS) is the most common adult-onset motor neuron disease, characterized by rapid loss of upper and lower motor neurons resulting in patient death from respiratory failure within 3-5 years of initial symptom onset. Although at least 30 genes of major effect have been reported, the pathobiology of ALS is not well understood. Compounding this is the lack of a reliable laboratory test which can accurately diagnose this rapidly deteriorating disease. This barrier has created a severe road-block for researchers trying to develop novel therapeutics and clinicians initiating treatments. Therefore, alongside identification of etiologies and therapeutics, development of a diagnostic and prognostic biomarker is a foremost research priority. Herein, we report the use of graphene, a sensitive nanomaterial, in combination with Raman spectroscopy to identify specific signals in the cerebrospinal fluids (CSF) of ALS patients which in the future can be used to develop a diagnostic biomarker. We demonstrate that the second-order overtone of in-plane phonon vibration energies (2D) of interfaced graphene can be sensitively modified by the components in the CSF. The n-doping mechanism employed by our platform is disease specific, with a 2D vibration energy shift of approximately 3.3 ± 0.3 cm^{-1} in the case of ALS patients.

538632 Xeno Nucleic Acids for Enhancing the Optical Stability of Nanosensors

Abstract:

The unique optoelectronic properties of single-walled carbon nanotubes (SWCNTs) motivate their use as next-generation optical sensors. The intrinsic SWCNT near-infrared fluorescence offers many advantages in terms of bio-transparency, sensitivity, and photo-stability. Characteristics required for *in vivo* continuous sensing, such as selectivity, biocompatibility, and solubility, can be tuned by suspending SWCNTs with biomolecules such as single-stranded DNA (ssDNA). Despite the ubiquitous use of DNA-wrapped SWCNTs for a variety of *in vivo* sensing applications, their behaviour in the presence of varying concentrations of the different ions present in biological media has yet to be studied.

In this work, we present a systematic exploration of the effects of various cations and anions on the fluorescence of ssDNA-wrapped SWCNTs. The presence of multivalent cations was found to induce fluorescence shifting of near-infrared fluorescence through the transient formation of a metastable state. The incorporation of artificial oligonucleotides, or xeno nucleic acids, was shown to mitigate the effects of ion-induced conformal drift through improved conformational resiliency. The synthetic biology approach presented herein offers a powerful demonstration for enhancing sensor selectivity and stability while enabling a previously unexplored space for engineering oligonucleotide-based SWCNT sensors.

538009 Quantification of Inflammatory Response and Morphological Change of SIM-A9 Microglia By Neuro-Probes

Abstract:

Fluorescent probes such as calcium dyes,[1] neurotransmitter nanosensors,[2] and voltage probes,[3] are utilized ubiquitously in neuroscience to study brain structure and neurochemical activity. While most of these probes target the study of neurons, their effects on another major cell type of the brain, microglia, are often overlooked. Microglia are the resident immune cells of the brain that number in the billions, and are central in maintaining the homeostasis of the neuronal environment through states of activation and quiescence.[4] In the brain, microglia mediate inflammatory responses to foreign and pathogenic material, which can elicit downstream neurotoxicity and neurodegeneration. Herein, we present a study on the effects of several brain chemistry probes on a SIM-A9 mouse microglial cell line. We show that single-walled carbon nanotube (SWNT)-based dopamine nanosensors are uptaken by, but are not cytotoxic to, SIM-A9 mouse microglia. The activity of SIM-A9 mouse microglia is next quantified with quantitative PCR and with an immune proteomics screen to monitor mRNA and protein expression and secretion levels of common inflammatory cytokine biomarkers. We show that SIM-A9 co-incubation with dopamine nanosensors induces a slight increase in expression of inflammatory cytokines TNF-α, iNOS, IL-1β, and IL-6. The observed nanosensor-induced cytokine upregulation is significantly lower than that caused by lipopolysaccharide stimulation. Live-cell imaging results further reveal that SIM-A9 interactions with SWNT nanosensors elicit a significant change in SIM-A9 cell morphology towards an elongated and ramified state, an *in vitro* indicator of a quiescent rather than inflammatory microglial state that halts cell proliferation. These contrasting effects indicate that nanosensors affect multiple microglial signaling pathways, possibly through a non-canonical immune response. We also discuss how common passivation techniques, namely PEGylation, can mitigate microglial inflammatory responses to brain imaging probes. PEGylated nanosensors show lower uptake into SIM-A9 cells, a reduced inflammatory response, and partial mitigation of SIM-A9 morphological change. Our results motivate greater attention to the effects of ubiquitously utilized neuro-imaging probes on the brain microglia.

References

1. Grienberger, C. & Konnerth, A. Imaging Calcium in Neurons. *Neuron* **73,** 862–885 (2012).
2. Beyene, A. G., McFarlane, I. R., Pinals, R. L. & Landry, M. P. Stochastic Simulation of Dopamine Neuromodulation for Implementation of Fluorescent Neurochemical Probes in the Striatal Extracellular Space. *ACS Chem. Neurosci.* **8,** 2275–2289 (2017).
3. Peterka, D. S., Takahashi, H. & Yuste, R. Imaging Voltage in Neurons. *Neuron* **69,** 9–21 (2011).
4. Kierdorf, K. & Prinz, M. Factors regulating microglia activation. *Front. Cell. Neurosci.* **7,** 1–8 (2013).

538299 Aggregation State Determines Uptake, Intracellular Processing, and Long-Term Fate of Single-Walled Carbon Nanotubes in Mammalian Cells

Abstract:

The intrinsic fluorescence of single-walled carbon nanotubes (SWCNTs), which exhibits exceptional photostability, near-infrared (NIR) tissue-penetrating emission, and microenvironmental sensitivity, makes them ideal candidates for a variety of biomedical imaging and sensing applications. Single-stranded DNA was found to disperse SWCNTs in water and enable their uptake into mammalian cells through an energy-dependent endocytosis mechanism. Previous empirical models of cellular uptake have been introduced that assume single SWCNTs are interacting with the cell membrane. Here, we show that DNA-wrapped SWCNTs reversibly aggregate in the presence of media containing serum proteins and this loose aggregation significantly affects the manner in which the SWCNTs interact with mammalian cells. We show that SWCNT aggregation not only alters the rate of uptake into the cells but further affects the endosomal processing and expulsion of the internalized material. These effects are highly cell-type dependent. Future work includes the development of a cellular uptake and expulsion model that accounts for the aggregation state of SWCNTs.

536164 Semi-Rational Design of Steroid Biosensors Using Compositionally Controlled Corona Phase Molecular Recognition: Pathway Towards In Vivo Monitoring

Abstract:

Steroid hormones contribute to numerous biochemical pathways controlling various physiological and pathological processes. Measurements of these molecules form the basis of many medical diagnoses; such as cortisol for Cushingâ €™s disease and mental illnesses; progesterone for female fertility; testosterone for androgenic traits; and aldosterone for adrenal cancer and irregular blood pressure. Moreover, these molecules are commonly administered through various routes as therapeutics. As such, real-time, continuous measurements of these steroids may provide new biomedical insights, enable new medical diagnoses, and become an integral part of therapy. In this study, we applied near-infrared, fluorescent single walled carbon nanotubes (SWNT) non-covalently functionalized with custom-synthesized polymers for steroid hormone detection. A novel steroid templating strategy was employed in the semi-rational design of the polymers. Thirty polymers varying in monomer composition and length were synthesized and interfaced with SWNT to create a sensor library. The sensor library was screened against a panel of 9 steroid hormones selected for their biological significance. Design principles were established by tracking sensor response against polymer composition. Two constructs selective for cortisol and progesterone were discovered. The progesterone sensor was encapsulated in a biocompatible hydrogel and shown to function in a biological environment. Preliminary experiments in rodents demonstrate the potential for *in vivo* monitoring.

535104 Substrate Functionalized Carbon Nanotubes As a Modular Tool for Tracking Soil Enzyme Activity

Abstract:

Soil microbes form an ecosystem that is vital for sustenance of plant, animal, and ultimately human life through the regulation of biological carbon, nitrogen, and phosphorous. The cycling of biological elements is largely regulated through activities of hydrolytic enzymes, such as proteases, glucanases, ureases, phosphatases, and sulphatases, and oxidative enzymes, such as phenol oxidases and peroxidases. While development of fluorophore-labeled substrates and fluorogenic substrate analogs has improved the ability to detect and characterize these enzymes, the cost of each of these custom reagents can limit their use. Moreover, they have limitations in the time required to complete the assay, and there are concerns about their use in complex mixtures (like soils). Single-walled carbon nanotubes non-covalently functionalized with substrates to enzymes may be used to track enzymatic activity. Theses sensors feature the advantages of simple preparation techniques, fast and facile detection procedures, and high modularity. Substrate degradation is directly tracked by monitoring the nanotube fluorescence, which decreases due to proximity quenching and changes in the local dielectric environment caused by substrate removal. The ability to use native substrates with little to no synthetic modification allows for order of magnitude reduction of costs. Tests with soil samples have demonstrated these sensorsâ€™ ability to respond to activity of proteases, chitosanases, and cellulases. This method may be further developed to detect a wider range of enzymes and adapted to distinguish between free soil enzymes, soil-stabilized enzymes, intercellular enzymes, and intracellular enzymes.

532216 Evolution of Nanoparticle-Based Synthetic Molecular Recognition

Abstract:

Molecular recognition is fundamental to the design of therapeutics, diagnostics, and sensors. Many of these technologies rely on natural molecular recognition elements such as antibodies, which have evolved to be highly selective for their molecular targets. However, despite their exquisite selectivity, antibody-based molecular recognition suffers from limited stability outside narrow ranges of temperature, pH, and salt concentration, and have limited *ex-vivo* stability. Conversely, synthetic molecular recognition has enabled detection of molecular targets over a broad range of extraphysiological conditions with varying success in selective and sensitive detection of target analytes. Recent developments in synthetic molecular recognition include synthetic antibodies based on electrostatic pinning of bio-mimetic polymers to the surface of single-walled carbon nanotubes (SWNT), whereby the polymer serves as the molecular recognition element, and the SWNT provides a near-infrared fluorescent signal in response to analyte binding (1-3). In this and other embodiments of nanosensor platforms, the principal challenge lies in the laborious process of identifying synthetically-produced selective and sensitive molecular recognition elements. Herein, we describe a polymer evolution-based platform, in which 10^12 unique SWNT-pinned polynucleotide polymers can be screened for their ability to bind a target analyte and provide a selective near-infrared fluorescence signal. Iterative selection of analyte-binding polymers that form a SWNT-surface-adsorbed phase for target recognition are identified through ionic desorption of sub-optimal polymers, and exponential amplification of synthetic polymers that recognize the target analyte. Next-generation sequencing identifies SWNT-polynucleotide conjugates selective and sensitive for the desired analyte. We demonstrate the utility of this platform for the evolution of a SWNT-polymer conjugate for neuromodulator serotonin, which we show to be reversible, to have a detection limit of 100 nM, and to have a dynamic range appropriate for *in vivo* imaging of serotonergic neuromodulation. Our platform is fundamentally generic in enabling the evolution of molecular nanosensors from synthetic polymer-nanoparticle conjugates for any desired analyte.

(1) Smith et al., Nat. Nanotechnol. 2009, 4, (11), 710-711.

(2) Zhang et al., Nat. Nanotechnol. 2013, 8, (12), 959-968.

(3) Kruss, S. et al., J. Am. Chem. Soc. 2013, 136, (2), 713-724.

535612 Development of Hydrogel Encapsulated Carbon Nanotube Based Biomonitoring System and Its Applications Toward the Detection of Riboflavin Administration

Abstract:

The development of carbon nanotube based sensors, where amphiphilic polymers on the nanotube surface screen the interaction between analytes and the nanotube in an effect known Corona Phase Molecular Recognition (CoPhMoRe), provides an opportunity for long term *in vivo* biomonitoring of analytes due to their near-infrared fluorescence, lack of photobleaching, and resistance to degradation. In this work, we develop a platform to track an intensity modulated sensor response to the administration of riboflavin in mice. A hydrogel encapsulation method allows for the diffusion of small molecules, such as riboflavin, to reach the corona of the nanotube sensors, while maintaining the sensors in a localized patch for repeated imaging. A wide field of view imaging setup allows for spatiotemporal imaging of the hydrogel patches. Furthermore, a reference sensor that is invariant to the analyte was developed and used to account for imaging artifacts from breathing and movement to facilitate the biomonitoring of riboflavin.

AUTHOR INDEX

AUTHOR INDEX

AUTHOR INDEX

AUTHOR INDEX

AUTHOR INDEX

AUTHOR INDEX

AUTHOR INDEX

AUTHOR INDEX